Digital Literacy and Inclusion

Danica Radovanović
Editor

Digital Literacy and Inclusion

Stories, Platforms, Communities

Springer

Editor
Danica Radovanović
Department of Technology Systems
University of Oslo
Oslo, Norway

ISBN 978-3-031-30807-9 ISBN 978-3-031-30808-6 (eBook)
https://doi.org/10.1007/978-3-031-30808-6

This Springer imprint is published by the registered company Springer Nature Switzerland AG
The registered company address is: Gewerbestrasse 11, 6330 Cham, Switzerland

Paper in this product is recyclable.

To my parents

Foreword

It is often thought that digital literacy, and its precursor, information literacy, originated with the widespread use of the digital computer. In fact, the issue of enabling good access to information, by providing finding aids, indexes, summaries, and the like, and ensuring that people can make effective use of them, goes back many centuries [see, for example, Blair (2010), Dover (2021), Duncan (2021), and Eisenstein (2005)]. Nor are the modern formulations solely restricted to computerised information, as witness Paul Gilster's notes on the value of printed materials in his classic formulation of digital literacy (1997). Similarly, the genesis of modern concerns about digital divides, information poverty, digital inclusion, and so on can be discerned well before the computer age, the nineteenth-century enthusiasm for public libraries, mechanics institutes, and self-improvement generally being only one obvious example.

However, as we move into the era dominated by what Luciano Floridi (2014) terms the *infosphere* and Caleb Scharf (2021) the *dataome*, we have a new, and even stronger, motivation to readdress these issues. When society worldwide is not merely supported by, but wholly dependent on, and largely shaped by information and communication technologies and the associated stores of information and data, the need to maximise digital literacy and digital inclusion is clearly crucial. Disinformation, hate speech, and the rest of social media's problems, the risks posed by thoughtless application of algorithms whose detailed workings are not understood, and the problems likely to be caused by AI deep fakes, are merely some of the symptoms. Moreover, it is clear that digital literacy, or the lack of it, has a direct influence on such pressing current issues as climate change, and vaccination against pandemics (Haider and Sundin 2022).

We now know that the encouragement and promotion of digital literacy, or digital fluency if one prefers, and of digital inclusion, is much more complex, and the approaches must be much more nuanced, multifaceted, and contextualised than has sometimes been assumed. The contributions in this book reflect this, with an impressively wide range of perspectives and research methods and an extensive geographical reach. The specific topics dealt with here, encompassing inter alia digital citizenship, smart cities, a variety of digital divides, critical thinking, algorithms, sustainable infrastructures, and health literacy, reflect the current breadth of the topic.

This book is clearly timely, and its editor has the expertise and experience to bring together a group of contributors able to encompass as much of these topics as is possible in a single volume. Its appearance is to be welcomed, and it deserves a broad readership.

City, University of London
London, UK

Lyn Robinson
David Bawden

References

Blair AM (2010) Too much to know: managing scholarly information before the modern age. Yale University Press, New Haven

Dover PM (2021) The information revolution in early modern Europe. Cambridge University Press, Cambridge

Duncan D (2021) Index, a history of the. Allen Lane, London

Eisenstein E (2005) The printing revolution in early modern Europe, 2nd edn. Cambridge University Press, Cambridge

Floridi L (2014) The fourth revolution: how the infosphere is shaping human reality. Oxford University Press, Oxford

Gilster P (1997) Digital literacy. Wiley, New York

Haider J, Sundin O (2022) Paradoxes of media and information literacy. The crisis of information. Routledge, London

Scharf C (2021) The ascent of information. Books, bits, genes, machines, and life's ultimate algorithm. Riverhead Books, New York

Acknowledgments

I want to acknowledge the support throughout the years of my postgraduate mentors and friends Lyn Robinson and David Bawden from the City University of London. Lyn and David, thank you for introducing me to information and media literacy 23 years ago before it became a widely researched and discussed topic. I am also grateful to Josef Noll, a colleague at the University of Oslo, thank you for introducing me to projects that make a difference on connecting the future. Massimo Ragnedda and Glenn Muschert, thank you for the early guidance on this book. I am grateful to Jonathan Muringani and J.D. Hildebrand for the final read, and valuable inputs. I appreciate the support and encouragement of the senior editor Mary James who helped me through the book creation process. Finally, I thank the contributors to this book for the kind and inspiring collaboration and to all the internet scholars, practitioners, and policymakers worldwide who advocate and act on the humanistic, ethical, and meaningful use and development of digital technologies.

Contents

About the Editor

Danica Radovanović, PhD, is a senior associate researcher at the University of Oslo (UiO) for digital social innovation and postgraduate curriculum development projects, and a Child Online Protection expert for the ITU (the International Telecommunication Union). Previously, she worked as a digital inclusion advisor at the Basic Internet Foundation, and as an information management specialist for the United Nations' FAO.

Danica is a PhD Chevening scholar at the Oxford Internet Institute (University of Oxford) and a doctorate graduate from the Faculty of Technical Sciences, Novi Sad. Combining both research and development, her work focuses on societal aspects of digital transformation and inclusion, digital literacy, and online safety. Danica has published two books and over 40 journal articles, book chapters, reviews, and essays.

About the Contributors

Asad Abbas is a research professor at the Writing Lab, Institute for the Future of Education, and a visiting professor in the School of Government and Public Transformation at Tecnologico de Monterrey. He received the Doctor of Management degree in Public Administration from the University of Science and Technology of China, Hefei. He is a member of the National System of Researchers in the National Council of Science and Technology of Mexico. He wrote several publications in international journals and proceedings and is reviewer of many international journals and conferences. His research focuses on public management and policy, strategic and innovation management, information systems, soft skills, and higher education.

Ahsan Ali is an associate professor in the School of Economics and Management at Zhejiang Sci-Tech University. Before this he was a postdoctoral research fellow at Tongji University. He has done his doctorate from the University of Science and Technology of China. His current focus of research is on exploring the implications of social media for organizations, and collective engagement of team members in knowledge management, leadership, and creativity. He has published several articles in reputed journals, such as *International Journal of Information Management*, *Journal of Organizational Behavior*, and *Information Systems Frontiers*.

Giovanni Boccia Artieri, PhD, is full professor of Sociology of Culture and Communication and head of the Department of Communication, Humanities and International Studies at the Urbino University Carlo Bo. His research is especially in communication and media studies, internet studies, innovation studies and technological change, and participatory and civic culture.

Hurriyet Babacan is a professorial research fellow at the University of Queensland's Australian Institute for Business and Economics, the research director at the Rural Economies Centre of Excellence, and a professorial fellow at James Cook University in The Cairns Institute. She has an extensive track record of leading multidisciplinary research in Australia and the Asia Pacific. She has published in national and international publications on economic and social development including publications for UNESCO. Hurriyet has received the Order of Australia, Prime Minister's Bi-Centennial Medal, and Telstra Business Women's Award.

Ataharul Chowdhury is an associate professor in the School of Environmental Design and Rural Development at the University of Guelph in Canada. Dr. Chowdhury has over a decade of teaching and research experience at universities, nonprofits, and private foundations in the Caribbean, Canada, Austria, and South Asia. He has collaborated with many leading international partners to help build more sustainable agriculture and rural development in remote and resource-poor communities around the world.

Leona Craffert (DPhil) is the director of the CoLab for eInclusion and Social Innovation at the University of the Western Cape, South Africa. She is a research psychologist with a special interest in the nexus between digital development and societal and institutional change, digital (social) inclusion, social innovation, and people development. Recent research interests include leadership for and in the digital economy, digital ethics, and multistakeholder partnerships for societal digital inclusion.

Allan Dale is a professor of Tropical Regional Development at The Cairns Institute, James Cook University, and the chief scientist for the Cooperative Research Centre for Developing Northern Australia, and a university fellow with Charles Darwin University's Northern Institute. Allan was previously the Chair of Regional Development Australia Far North Queensland and Torres Strait Inc. and the CEO of Terrain NRM. He has held executive and management roles within the Queensland Department of Natural Resources and Mines (General Manager – Strategic Policy) and the Department of Community Services.

Michael Dezuanni undertakes research about digital media, literacies, and learning in home, school, and community contexts. Professor Dezuanni is the program leader for Digital Inclusion and Participation for QUT's Digital Media Research Centre which produces world-leading research for a creative, inclusive, and fair digital media environment. He is a chief investigator in the ARC Centre of Excellence for the Digital Child. Michael has been a chief investigator on six ARC Linkage projects with a focus on digital literacy and learning at school, the use of digital games in the classroom, and digital inclusion in regional and rural Australia and in low-income families.

Uvasara Dissanayeke is a senior lecturer in the Department of Agricultural Extension at the University of Peradeniya, Sri Lanka. Dr. Dissanayeke has extensive experience studying agricultural digitalization in Sri Lanka and teaching courses on technology and development in conjunction with the Postgraduate Institute of Agriculture (PGIA). Her publications include studies on the use of m-Learning and social media in agriculture education.

Valeria Donato is a PhD student in Sociology of Communication at the Department of Communication Sciences, Humanities and International Studies of the University of Urbino Carlo Bo. Her main research interests revolve on Chinese digital diplomacy, Chinese strategic communication, and Chinese nation branding.

Nicholas Emtage is a business insight analyst at James Cook University who focuses on student success, retention, completions, and marketing. Prior to this Nicholas worked as a consultant and researcher specializing in socioeconomic research related to agriculture, natural resource management, and rural development. Nicholas's research work has been published extensively both nationally and internationally.

Talía González-Cacho received the master's degree from the Universal of Alcala, Madrid, Spain, on the Project of Architecture and City Design. She is currently Associate Professor in the School of Architecture, Art and Design at Tecnologico de Monterrey, Mexico. She holds the NOVUS funding for educational innovation research from Tecnologico de Monterrey since 2015. Her current research focuses on sensorial urbanism and higher education.

Gordon A. Gow is a professor cross-appointed with Sociology and Media and Technology Studies at the University of Alberta in Canada. Since 2012, Dr. Gow has been leading an ICT4D project on technology stewardship for agricultural communities of practice in Sri Lanka and the Caribbean. His approach to research emphasizes participatory methods for digital literacy development and community-based digital leadership initiatives.

Rachel Hay is a social scientist and researcher for the College of Business Law and Governance at James Cook University. Dr. Hay is a volunteer data analyst at the Better Internet for Rural Regional and Remote Australia advocacy group where more than 4000 survey responses have been collected that help describe the connectivity literacy concept. She is a knowledge broker at the Tropical North Queensland Drought Hub and adoption stream leader in the JCU Agricultural Technology and Adoption Centre (Ag-TAC). Rachel is a research fellow with JCU's Cairns Institute and a member of the Centre for Tropical Environmental and Sustainability Science, and Pacific Connect.

Jason Hung is a final year PhD candidate in Sociology at the University of Cambridge. He is also a fellow at Harvard University Asia Centre, a visiting fellow at Academia Sinica, and a former Global Health PhD fellow at the United Nations University Institute for International Global Health.

Natasha Katunga (DPhil) is a researcher at Western Cape CoLab for eInclusion and Social Innovation, which is based at the University of the Western Cape, South Africa. Currently, her work focuses on digital inclusion, digital transformation, and impact assessment of digital inclusion interventions. Her research interests are in the ICT4D and social media for development space, particularly societal inclusion and digital social innovation.

Carlynn Keating (DPhil) is a researcher at the CoLab for eInclusion and Social Innovation at the University of the Western Cape, South Africa. Her research has primarily centered on digital inequalities, with a particular focus on intersections between gender and ICT. Her work at the CoLab includes monitoring and evaluation of digital inclusion interventions and education in the digital transformation domain.

Kristjan Kikerpill (PhD in Sociology) is lecturer in Information Law and Digital Sociology at the Institute of Social Studies (University of Tartu, Estonia). His main areas of interest and research are lying and online deceptions, social engineering, technology-mediated surveillance, and the social aspects of artificial intelligence.

Maximilian Krüger is a doctoral research associate at the Institute for Information Systems and New Media, University of Siegen. His doctoral work addresses the role of digital interventions in support of refugees and forced migrants' efforts to make themselves at home in Germany and the possibility and challenges of participation in and through design.

Amber Marshall is a postdoctoral research fellow at Queensland University of Technology's Digital Media Research Centre, and an adjunct research fellow at The Cairns Institute, James Cook University. Dr. Marshall's research focuses on digital inclusion and sustainable rural development. Her research interests include digital AgTech and data, digital inclusion ecosystems, remote telecommunications infrastructure (both technical and social), and digital skills and capability development. Amber principally employs ethnographic methods (co-design, participant observation, interviews, focus groups) to immerse herself in rural contexts and strives to develop research outputs that translate into actionable options for local stakeholders.

Juliana Novaes holds a master's degree in Science and Technology Studies from Maastricht University. Her research focuses on the intersections between technology and society, using computational methods to understand societal issues. She has been working on topics related to the digital divide in Latin America since 2019, having participated in the ITU's Development Sector as a member of the Brazilian delegation and as an Internet of Rights fellow at ARTICLE 19.

Danica Radovanović, PhD, is a senior associate researcher at the University of Oslo, Norway, for digital social innovation and postgraduate curriculum development projects, and a policy specialist at the Peace Research Institute Oslo's NORM project. Danica is a PhD Chevening scholar, Oxford Internet Institute (University of Oxford), and a doctorate graduate from the Faculty of Technical Sciences, Novi Sad. Combining both research and development, her work focuses on societal aspects of digital transformation and inclusion, digital literacy, and online safety. Danica has published two books and over 40 journal articles, book chapters, and essays.

Jeet Ramjattan is an experienced agricultural extension officer with the Ministry of Agriculture, Land, and Fisheries in Trinidad and Tobago, working with academic, government, and other organizations on farmer field schools in Trinidad. He is also a doctoral student in Agricultural Extension at the University of the West Indies (UWI) and has been an active collaborator in the technology stewardship training program.

Raquel Renno holds a PhD in Communication and Semiotics from PUC-Sao Paulo on the intersection between urban studies and the media environment. She is also digital program officer at ARTICLE 19, focusing on spectrum management issues, primarily in and around the ITU's Radiocommunication Sector. She has been working on topics related to human rights and the social impact of digital technologies since 2003. She also worked together with local grassroots communities in Central and South America on projects to tackle the digital divide among women and girls in familiar agriculture and indigenous communities.

Guillermina Benavides Rincón is professor and director of the Master's in Strategic Foresight at the School of Government and Public Transformation, Tecnologico de Monterrey. She holds a PhD in Social Work from the University of Texas at Arlington and the Universidad Autónoma de Nuevo León. She has participated in consulting projects advising the federal government (2006–2007) and state (2005–2006) and municipal governments (2007). Her research interests are futures studies, strategic foresight, inter-institutional coordination of public policy, and research methods. She has several publications in national and international journals and is part of the Editorial Board of the World Futures Review of SAGE publishing.

Kerry Russo is the Associate Dean, Learning and Teaching at James Cook University, College of Business, Law and Governance. Her role as Associate Dean includes leading the development of a learning and teaching culture committed to excellence, innovation, and a positive student learning experience. Kerry sits on numerous Australian national and state committees including Chair of the Queensland branch of the Higher Education Research Society Australia and Executive Committee Member of the Australian Business Deans Council, Teaching and Learning Network. Kerry is an experienced educator, whose social justice values demonstrate a commitment to equality in quality education.

Sushant Sonar is data visualization specialist at India Health Action Trust (UPTSU). He worked on multiple projects related to Digital Health Sector and Analytics with the aforesaid organization. He has contributed his insights on multiple projects namely UP Ke Swasthya Kendra, an application that gives a 360-degree view of a facility concerning HR, supply chain, programs, and infrastructure. He has over 11 years of experience in project management, process management, and IT solutions development, leading and executing initiatives in the IT services industry.

Ritu Srivastava has over 16 years of experience in the development sector focusing on the ICT and sustainable development of underprivileged communities/marginalized groups of society. Her research interest lies in areas of broadband policies, gender and access, gender and internet governance, violence against women, open spectrum policies, digital literacy and digital health, and community development. She has represented India for the United Nations Human Rights Council, United Nations Secretary General's High-Level Panel for Digital Cooperation, and organizational representative at the Global Network Initiative. She holds a Master's degree in Electronics and Telecommunication.

Leo Van Audenhove is professor and head of the Department of Communication Studies at Vrije Universiteit Brussel. He is a researcher at IMEC-SMIT-VUB and extraordinary professor at the University of Western Cape. In 2013, he was instrumental in setting up the Knowledge Center for Media Literacy in Flanders, of which he subsequently became the director. The center was established by the government as an independent center to promote media literacy in Flanders. His research focuses on internet governance, media literacy, e-inclusion, and ICT for development.

Minna Vigren is a digital media studies and science and technology studies scholar at Aalto University, Finland. The common thread of her research has been the question of human agency and power in the thoroughly digitalized society. She has a special interest in sociotechnical imaginaries and methods to foster imagination on alternative futures. Her work is interdisciplinary, social scientific, and qualitative in nature, with an interest in developing new methodologies and conducting research experiments. She is currently studying imaginaries of sustainable digital futures in her Academy of Finland postdoctoral research project.

Anne Weibert is a postdoctoral research associate at the Institute for Information Systems and New Media, University of Siegen. Her research interest is in computer-based collaborative project work and inherent processes of technology appropriation, intercultural learning, and community-building. She has conducted participatory design works with children and adults in socially and culturally diverse settings. She was awarded the "Förderpreis des Augsburger Wissenschaftspreises für Interkulturelle Studien" in 2007, and the "Rolf H. Brunswig-Promotionspreis" in 2020.

Johanna Ylipulli is a docent in digital culture and an anthropologist focusing on urban contexts and digitalization. She is leading an Academy of Finland Research Fellow project titled *Digital Inequality in Smart Cities* at Aalto University, Finland. Dr. Ylipulli's professional background is highly interdisciplinary: for the last 12 years, she has been conducting research on urban environments and on new technology with architects, designers, and engineers. She has worked within diverse fieldwork settings and participated in planning several design-related empirical studies. She is interested in investigating how critically oriented anthropological thinking can be successfully combined with design research and technology development.

Elisabetta Zurovac, PhD, is a researcher at the Department of Communication Sciences, Humanities and International Studies of the University of Urbino Carlo Bo. Her research interests concern digital media and the self-narrative practices connected to them, with a particular reference to visual data, generations, and screen cultures.

Chapter 1
Rethinking Digital Literacy

Danica Radovanović

In the past few decades, digital technologies rapidly occupied our analogue lives, and we started to shift more towards networked digital environments. In scholarly and industrial discourse, this socio-technological shift and process are referred to as digital transformation. The term digital transformation is used by the public to describe digitalization and moving from analogue platforms, services, communities, and social inclusion activities to digital ones. And it is more than that. The proliferation of digital technologies is assumed to accelerate the digital transformation to leverage opportunities and simultaneously address socio-economic challenges. The latest data from the ITU (The International Telecommunication Union 2022) indicates that approximately 5.3 billion people were using the Internet in 2022. At the same time, the excluded one-third of the world's population remains either offline or in the digital transformation process. This digital exclusion exacerbates digital divides based on mere internet access. Digital literacy plays a crucial role.

The relationship between digital inclusion and digital literacy is *symbiotic*, as digital transformation affects over half of the world's population and all areas of their everyday lives. However, this relationship is not *synergetic* as not everyone can access the internet and hundreds of millions lack digital literacy skills. In formulating the construct of digital literacy in the twenty-first century, we must leave behind the digital literacy definitions, frameworks, and practices of the past two decades.

This book is about new knowledge and insights on the ways digital literacy is being practiced, expressed, and promoted in various geographies and communities, and how addressing the digital skills gap can contribute to digital inclusion. Contributors to this book provide insights from the case studies, research, and recommendations about what is needed to reach a human-centric approach to digital transformation that would also benefit unconnected and underserved communities.

D. Radovanović (✉)
Department of Technology Systems, University of Oslo, Oslo, Norway
e-mail: danica@basicinternet.org

D. Radovanović (ed.), *Digital Literacy and Inclusion*,
https://doi.org/10.1007/978-3-031-30808-6_1

Moreover, they recognize that digital technologies advance at a pace that requires us to rethink and revise our understanding of digital literacy. In the early days of the internet, digital literacy could be a simple matter of operating a browser or formatting text in word processing software, or communicating via email. Web 3.0 and the anticipated fifth industrial revolution will require us to adopt a revised model of digital literacy that empowers users to understand, navigate, create, and collaborate through the digital world of artificial intelligence software, augmented reality, interaction with robots and algorithms, and different virtual platforms. In this rapidly evolving digital environment, we could say that digital literacy encompasses the set of capabilities and skills and values; it is the ability to mindfully analyze, process, design, and produce information; to develop and employ critical thinking skills in the landscape of mis- and disinformation practices at digital platforms; to create, collaborate, engage, and communicate with others in a respectful and meaningful way; to understand the algorithms' mechanisms and strategically interact with artificial intelligence and similar platforms; to use the internet in a responsible, safe, and ethical manner having in mind the data privacy and digital footprint; to be accountable and respectful for one's actions online, and to be able to the understand the consequences of one's behavior. These capabilities and skills are essential for creating value and for helping us become independent critical thinkers, contributors and creators, and resilient digital citizens.

Over the past 3 years of the global socio-political disruption, we have seen that the situation magnified digital divides and highlighted the need for digital inclusion. Some internet scholars (Ragnedda 2017; Radovanovic et al. 2020; Helsper 2021) have highlighted the three levels of the digital divide. Level one reflects the *connectivity to broadband internet.* Level two in *digital literacy and skills*. And the third level of the digital divide implies *life benefits and opportunities gained from internet access and obtained digital skills.*

The first level of the digital divide is the lack of *reliable, affordable, meaningful, and universal connectivity and the infrastructure* necessary to access the information. In other words—internet access should serve as an enabler and human right. However, almost 3 billion people are not included in the digital revolution and do not enjoy this right.

The second level of the digital divide relates to the *lack of digital literacy and digital skills*. Internet connectivity is not enough. The technological infrastructure and internet access are just starting points. Digital skills are the key factors for digital transformation and sustainable development (Carretero et al. 2016; van der Velden 2018; Radovanovic et al. 2020). We saw during the disruptive years of 2019–2022 how much and to what extent some individuals and communities lack the digital skills necessary to participate in the fast-paced digitalization of society. The same applies to many of us, even academics and IT professionals. At some time, each of us had to reach out and ask our junior colleagues to help to set up the technology and install applications. We had to learn—and someone had to teach us—how to record a video, create, and share content with a larger group of people, use the cloud, change privacy settings, change the background, and press "unmute." Digital skills are intangible, and, yet, like other forms of human capital, they are unequally valued, distributed, and used to extract capital or benefit (Radovanovic et al. 2015).

This brings us to the third level of the digital divide which implies *life benefits, improved livelihoods, and opportunities* gained from internet access, usage, and skills. The third level of the digital divide refers to the unequal distribution of life opportunities: economic, social, cultural and personal well-being benefits (Helsper 2021), such as attending school, graduating, getting a better job, and participating in social, economic, and cultural life. Even internet access and digital skills do not guarantee that individuals will seize these affordabilities and use them productively—variations on the individual level influence the equality of the seized benefits. Also, socially excluded groups and vulnerable and disadvantaged communities (e.g., elders, low-income earners, and people with a disabilities) are less likely to have affordable internet access or the necessary skills to use connections effectively.

Before COVID-19, children, women, and vulnerable and marginalized groups were least likely to have access, skills, and benefits from the use of digital technology. According to a recent report of the Alliance for Affordable Internet, the digital gender gap is substantial as men are 21% more likely to be online than women globally, rising to 52% in the least developed countries. The ITU's regional estimates for Africa put the gender ratio at nearly three-to-two in favor of men over women. According to the GSMA, about 234 million fewer women in low- and middle-income countries use the mobile internet than men. This divide is stark in sub-Saharan Africa and South Asia, where the gender gap persists over 55% more men than women. This dire disadvantage has been exacerbated due to the lack of connectivity and digital skills, making the digital divide even more alarming.

All three levels of the digital divide are more present in under-served, rural, remote areas and the Global South. Undoubtedly, digital inclusion is one of the key drivers for bridging the digital divide. It poses opportunities and challenges to the social, economic, cultural, and environmental sustainability of individuals, organizations, communities, and society (Muringani and Noll 2021). These existing and emerging challenges have given birth to new types of opportunities in the form of values and digital competencies. Digital resilience is one of the values. Digital resilience is the ability to use technological tools to strengthen ourselves and our knowledge on an individual and community level (Van Der Velden 2021). As a result, we can successfully and safely navigate online, deploy critical thinking skills, engage and interact with others, and learn and relearn (Haythornthwaite and Andrews 2011; Radovanovic et al. 2015). It is a valuable asset for navigation, prosperity, and sustainability in the digital age, especially during turbulent times.

As for digital competencies, digital literacy is central to building resilient communities. Yet, as we've seen, almost three billion people cannot access digital technologies or use them, and consequently cannot access economic resources and opportunities that come with an internet connection and digital skills.

In light of these opportunities and challenges, this book presents a palette of stories, case studies, theoretical and research implications, and best practices on digital literacy as a prerequisite for effective digital inclusion.

Stories, Platforms, and Communities

Stories in this book explore digital literacy and inclusion in different domains of everyday life in various worldwide geographies. Digital literacy and digital inclusion are explored in three main sections: theoretical implications, digital literacy textures through the filter of education, and convergent practices that showcase the digital literacy initiatives applied through communities of practice around the globe.

The contributions in this book provide myriad perspectives, stories, research methods, and extensive geographical reach. The chapters cover various forms and modalities of digital literacies, digital citizenship, smart cities, a variety of digital divides, critical thinking, algorithms, community networks, and sustainable infrastructures.

The first section of the book sets up the context and the theoretical understanding related to digital literacy, digital inclusion, and digital citizenship. The complexity and depth of societal development entangled with digital technologies challenge the prevailing conceptualizations of digital literacy. Johanna Ylipulli and Minna Vigren, in the chapter *From skilled users to critical citizens? Imagining and future-making as part of digital citizenship* explores how the notion of *digital citizenship* could be complemented and expanded to include an *ability to imagine alternative future trajectories*. They provide theoretical insights supported by examples from their recent empirical studies while focusing on experiences and perspectives of individuals' understanding of how people perceive their technological agency—or the lack of it. Authors draw from approaches provided by *design-oriented thinking*, especially *speculative design* and *Participatory Design (PD)*, to broaden the discussions linked to digital literacy and digital citizenship towards understanding *active, participatory future-making as a means to increase technological agency and awareness*. Design as a future-oriented field offers both ways to conceptualize our relationship with change and beneficial practices for democratizing future-making and fostering creativity. They underline that it is not just the individual's responsibility to become a 'proper citizen' in the digital society, but also the society is accountable for arranging adequate conditions in which critical, transformative, and resilient digital citizenship can flourish.

We continue with the digital literacy, fostering creativity, and digital citizenship theme in the next chapter. The chapter offers insights into how creative work can be a means to the building of *urban data literacy*—an ability to make sense of phenomena in the urban sphere by using and interpreting data and information. In *Sensing the city: a creative data literacy perspective,* Anne Weibert and Maximilian Krüger explore from a digital citizenship perspective the need for literacy that can navigate the manifold kinds of data surrounding us in everyday city life (Cowley et al. 2018), understand their potential effects and make informed choices about data (Kunze 2020). Focusing on experiences from a series of workshops in a midsize city in Germany, this chapter argues for the inclusion of making and crafting as alternate methods for urban data literacy. They show that these can be a means to bring unseen city life dimensions to the fore and to include youth and those whose access to the digital sphere is challenged in the discourse on data and its implications for everyday city life. The tactile and artistic approach to data, coming out from these workshops and its meaning in the city, enables their discussion across communities and adds to the data basis used to legitimize urban design choices and foster a broader degree of participation.

With the accelerated use of digital media platforms, cyber threats such as phishing, identity theft, and data privacy breaches (among others) are simultaneously emerging. The need for strategic and regulatory direction regarding online safety and cybersecurity preparedness is crucial. Particularly, one of the most common forms of social engineering attacks in digital environments is broadly referred to as "phishing". In the chapter *Scanning for scams: Local, supra-national and global events as salient contexts for online fraud,* Kristjan Kikerpill uses the *mazephishing* framework to explain how digital literacy instruction can benefit from observing the way in which salient social circumstances create a fertile ground for disseminating online scams. The *mazephishing* framework comprises three primary components: the social context from which specific scam messages obtain their salience, the media or channels used to circulate the scam messages, and the influencing techniques employed in the actual scam messages. The chapter presents three real-life examples for different levels of salient contexts, that is, local, supra-national, and global contexts: the tax season in Estonia and the United States as an enabler of related online scams, supra-national level commercial events, that is, Amazon Prime Day and Black Friday, as well as the global microchip shortage that drives international gaming console scams. The *mazephishing* framework acknowledges that not all events are of equal importance for different societies around the world and that even global events are lived and experienced differently across the globe. Thus, the framework provides digital and media literacy instructors with a valuable tool and set of principles for analyzing how events occurring on different scales are used to exploit scam recipients' vulnerabilities.

We move across the continent to Southeast Asia (SEA), where digital growth is evident in the region. Despite incorporating technologies in the development of smart cities, we still face a stark digital divide and digital exclusion suffered by digital (semi-)illiterate, less-educated, and economically underprivileged individuals and communities. Jason Hung, in the chapter *How Southeast Asian countries can better arrange and deliver internet policies to defy the digital divide* provides a systematic overview and analysis of how countries digitally harness and maximize the benefits from online platforms. He discusses how the digital divide, marginalization, and exclusion as processes continue to be ingrained in the SEA region, urging a need for intra- and inter-countries' inclusive responses to e-development. The author raises concern over digital transformation and explores how different SEA stakeholders can respond to these risks.

The second section of the book reveals textures and stories related to *digital literacy and education.* The digital divide affects the learning and education ecosystem and is significantly magnified during the COVID-19 crisis. Students' inability to develop effective digital skills and access education and quality information can result in digital exclusion. These socio-technological changes bring challenges and implications for the e-learning and education ecosystem. As the schools were shut down, the government opted to replace in-class learning with various means of distance education. This further revealed the dimensions of the digital divide at a larger scale to the concern of educators and policymakers. Kerry Russo and Nicholas Emtage, in *the Digital divide and Higher education,* researched how the pivot to online learning during COVID-19 has broadened the digital divide in the Australian higher education sector. Many students from disadvantaged backgrounds entering university were grappling with the necessary digital skills required to participate in a digital learning setting. Conceptualizing the growing inequalities arising from a widening digital divide, this chapter investigates the impacts of a digital divide on the university student experience. Using a quantitative approach, the chapter analyzes the digital divide in Australian higher education. Empirical data is examined to determine a link between the students' self-reported digital skills, prior digital experience and preparedness for university study with access to digital resources, demographic factors and geographic location. This chapter highlights the relationship between access to a learning management system (LMS) or digital curriculum during secondary school and digital literacy. Students immersed in digital learning environments during secondary schooling were more likely to be digitally literate for university study. Disadvantage indicators, prior digital experience, and digital literacy are examined to provide insight into the new barrier in higher education, the digital learning environment.

Critical thinking skills are one of the crucial digital skills in today's information age. We face the pervasiveness of misinformation and disinformation on social media. Equally important, digital critical thinking skills are relevant in social processes of collaboration and engagement in education and workforce dynamics. In *Students' use of social media and critical thinking: The mediating effect of engagement,* Abbas, González-Cacho, Radovanović, Ali, and Rincón, empirically explore the mediating role of students' social media engagement, and their ability to think critically. To achieve the aim of the study, the authors designed an online survey with questions related to the use of social media, engagement, and critical thinking through the deployment of digital literacy skills. Results from the data analysis support all proposed hypotheses and affirm that engagement is partially mediated between the use of social media and the critical thinking skills of undergraduate students. The findings confirm that using social media-based course activities is helpful for university students to engage with other peers by deploying digital literacy skills to analyze, share, and communicate relevant information and knowledge about specific topics within the relevant course structure.

The next chapter deploys digital ethnography to analyze the relationship between the digital skills and algorithmic practices enacted by young users and teenagers on TikTok, a popular short-form video hosting platform. In *Tales of visibility in TikTok: The algorithmic imaginary and digital skills in young users*, Zurovac, Artieri, and Donato, research TikTok's algorithmic awareness and practices. The algorithmic awareness seems crucial (Karizat et al. 2021) to the user's experience within TikTok: it allows the development of algorithmic folk theories (Eslami et al. 2016) that lets users behave, engage, and also resist the platform's structure. To gain visibility or to elude the attentive scrutiny the algorithm performs, a user should develop technical knowledge and digital skills. The online ethnographic method enabled the authors to understand in-depth how young users (a) learn about the algorithm and (b) develop or employ digital skills which enable them to (c) perform leverage tactics in response to the platform's logic. These dimensions were then qualitatively assessed to understand how peculiar digital skills (social interaction skills, content creation skills, and ethical behaviour online) are employed in relation to the algorithmic imagery. The discussion takes place, within the chapter, through the analysis of three case studies: #traumatok, Breonna Taylor trend, and teenagers' practices on LIVE streaming.

The third section of the book brings the *convergent practices and hands-on approach to the inclusion of digital literacy in the communities of practice*. It opens with two chapters related to digital inclusion initiatives in the rural and agricultural sectors, first in Global South, (Sri Lanka, Trinidad), and second, in rural and remote areas in Australia. The agriculture sector in the Global South is undergoing profound changes because of digital transformation. There are concerns these efforts will advantage global agribusiness while marginalizing smallholders and disrupting long-established labor patterns and social relations, especially for women and youth. Nonetheless, digital technology also promises to help these farming communities to maintain their independence and enhance their livelihoods. In *Digital Literacy and Agricultural Extension in the Global South,* Gow, Dissanayake, Chowdhury, and Ramjattan introduce and explain how an interactionist view on digital literacy can contribute to this objective when combined with a capabilities-centric approach to human development. An interactionist approach to digital literacy is important to consider with communities of practice to facilitate new practices involving unfamiliar ICTs. They provide a conceptual framework that connects the literature on agriculture innovation systems with an interactionist view of digital literacy from organizational studies. The authors explain how this view of digital literacy aligns with a capabilities-centric approach within ICT for development (ICT4D) by situating it along four degrees of empowerment from Kleine's Choice Framework (Kleine 2013). At a practical level, they offer best practices on how agriculture extension and advisory services (EAS) can promote digital self-determination for farmers and other agriculture workers. However, they suggest that EAS organizations must develop digital literacy strategies that will empower smallholders to make informed choices about new ICTs and how they will be integrated into work practices and community life. The final section of the chapter includes examples of how their conceptual framework has been applied in an ongoing action research study and technology stewardship training program involving EAS practitioners in Trinidad and Sri Lanka.

Low levels of meaningful connectivity and digital inclusion present enduring challenges for rural Australian communities and industries. Not only is there a historical lack of telecommunications infrastructure in rural areas, but rural people—including farmers—tend to have fewer digital literacies and skills to use digital connections and technologies effectively. Additionally, the recent COVID-19 disruption has highlighted and exacerbated the inherent social and economic disadvantage of people who lack the means to access and use digital technologies reliably (Thomas et al. 2020), such as those in rural Australia. In *Connectivity literacy for digital inclusion in rural Australia*, Marshall, Hay, Dale, Babacan, and Dezuanni, undertook a qualitative study in the State of Queensland in Australia that aimed to investigate underlying factors of low levels of digital inclusion in rural households and communities. The authors uncovered a set of essential skills for digital inclusion that did not fit neatly into the three-pillared framework they took into the study. They used qualitative methods (ethnographic interviews, focus groups, and participant observation) to give a voice to members of one of Australia's least digitally included populations. While the authors gathered insights into farmers' challenges associated with access, affordability, and digital ability, they also observed how people needed to acquire *connectivity literacy* (a concept originating from industry but not yet investigated by scholars) to achieve digital inclusion. The chapter sheds light on the complexities of connecting to telecommunications services in rural Australia. It described connectivity literacy as principally about setting up local hardware and networks, responding to technical outages when they occur, and navigating a complex consumer environment to ascertain the best connectivity options.

Digital literacy has been on the agenda of different international bodies and governments during the last decades, especially in the Global South. Despite many efforts, low digital literacy is one of the main reasons for developing countries' digital divide. Latin America presents a combination of low scores in international education rankings combined with a constant increase in the adoption of digital technologies for communication (International Telecommunication Union 2021; GSMA 2021). At the same time, countries from the region share a history of popular education as a model for teaching and learning media and communication. In the chapter *Community networks as sustainable infrastructure for digital skills,* Raquel Rennó and Juliana Novaes conducted a qualitative study on community digital literacy projects and telecommunication providers in rural areas of Latin America. They aim to find out if they could offer alternative educational methods that would respect their cultural specificities. The authors carried out three case studies in remote and rural areas in Brazil and Mexico, trying to find out lessons that could be learned from those initiatives. The analysis of the collective case studies showed that locally developed Internet service providers (ISPs), such as community networks, are more familiar with the community's context and interests and can produce practical educational tools and content. All three case studies have a predominance of onsite training and provide a bottom-up approach to digital skills learning in spaces where megaprojects from national and international commercial agreements have disrupted the environment and social tissue of local communities. They are essential in offering capacity-building initiatives for communities living in unconnected and underserved areas to increase digital literacy. Given that community ISPs are managed and maintained by the community, they are also cognizant of the socio-economic and cultural differences of the community, allowing them to embrace digital technologies within their culture, combining oral tradition with new media environments.

Speaking of underserved areas and vulnerable and digitally excluded communities, there has been a significant focus on digital skills development in the South African government's embrace of digital transformation. Community organizations that have integrated digital inclusion offerings as part of their services have been instrumental in providing digital skills training and alternative learning options for people unable to afford traditional education institutions. Unfortunately, limited information is available as to the outcomes for intended beneficiaries, as well as recommendations towards assessing such outcomes in under-resourced communities (URC). In *Digital inclusion interventions for digital skills education: Evaluating the outcomes in semi-urban communities in South Africa*, authors Katunga, Keating, Craffert, and Van Audenhove present the findings of a quantitative survey study that sought to contribute to the practice of assessing the outcomes of digital skills training interventions. The chapter provides insight into meaningful benefits derived by beneficiaries of a mobile (digital) literacy training course; salient factors contributing to such outcomes; and the application of methodological approaches and processes to evaluate the outcome of digital skills interventions in URC. The Mobile Literacy course was developed to enable participants to master the digital literacy competencies, navigating information, communication and collaboration, safety and security, and problem-solving and transacting. The course was implemented in 2020 through four community-based organizations in different peri-urban and rural environments in the Western Cape province of South Africa. The authors used a quantitative research approach, applying survey methodology to assess the outcomes of the Mobile Literacy training intervention. It is envisaged that the evidence-based insight emerging from this study allows for a more nuanced understanding of meaningful outcomes that may transpire from digital inclusion interventions. It also informs and encourages the effective practice of digital skills and the assessment of interventions aimed at improving them, particularly in URC.

And finally, the unavoidable topic of the past couple of years of the global disruption and COVID-19 that have affected public health the most. Digital technologies across the globe are being harnessed to support public health by identifying new cases, contact tracing, surveilling patients, and evaluating interventions based on mobility data and communication with the public. The pandemic raised some immediate questions about the capacity and access of the existing public healthcare system and government relief measures. However, the affluent sections of society had access to smart devices and internet connections and could shift to online mode to access their basic needs and health services. In contrast, the other section of society struggled to make ends meet with no hope in sight. Harnessing the use of digital technologies, digital health literacy (DHL) became an indispensable tool for the health workforce, including frontline workers, to provide primary health services during COVID-19. Having a digitally literate health workforce is an essential element for the success of establishing a health ecosystem. Ritu Srivastava and Sushant Sonar, in *Digital health literacy: a prerequisite competency for the health workforce to improve health indicators in times of COVID*, analyze the case study of Uttar Pradesh, with a population size of 200 million. Uttar Pradesh became the first state in India to develop an integrated and unified COVID-19 mobile platform. It ensured that every citizen of the state is tracked, tested, and treated across the continuum of care—bringing together all COVID health facilities, laboratories, state, district, and field staff onto a common platform. The chapter shows how digital health literacy has been imparted to 200,000 health officials and frontline workers across the state's 75 districts and 59,163 village councils. The authors measured the complete spectrum of DHL 2.0 skills from searching, selecting, appraising, and applying online health information and healthcare services provided at different healthcare facilities. The chapter aims to establish digital health literacy as a prerequisite requirement for health professionals and the workforce to effectively deliver healthcare services, specifically in the public health sector. The authors argue that digital health literacy skills are mandatory to ensure the health workforce and practitioners can integrate their knowledge and digital health literacy skills into optimal health behavior. They offer lessons learned from the case study and further recommendations for decision-makers in the field of digital health literacy.

Digital Literacy's Role in Our Digital Future

It is more than ever relevant to address the present digital transformation challenges in society to understand, regenerate, renew, and strive in the (digital)future. In this book and individual chapters' contributions, we have seen that digital transformation encompasses individuals, communities, and marginalized groups having access and digital skills to use internet technologies and, therefore, to participate meaningfully and benefit from today's growing information society. And it is 2023, and in the current digital landscape, the revolution and revelation are in the online sphere: online work, online socialization, online learning, e-commerce, e-government, and all that on the remote. In this landscape, digital literacy plays a crucial empowering and enabling role. COVID-19 showed the necessity for digital literacy development to facilitate internet users' access to online platforms, services, and information. And the chapters that follow in this book include various aspects and modalities of digital literacy: digital skills related to creativity, urban data literacy, digital citizenship skills, digital literacy in education, connectivity literacy, online safety skills, problem-solving and critical thinking digital skills, data literacy skills, mobile digital literacy, algorithmic digital skills, digital health skills, etc.

To take advantage of the benefits of digital transformation, we need an ongoing dialectic on socio-technological, cultural, educational, and economic participation in the online world and exploration of the gaps that hinder participation. On the other side, we have emerging technologies such as artificial intelligence, augmented reality, and automation. The growth of these technologies raises important questions about our current legal systems, policies, and advocacy strategies and how they can mitigate the human rights risks that may be affected by these technologies. Technology needs to help us make better decisions and improve our livelihoods, especially in the other half of the world that is not connected to the internet. The book highlights important nuances and raises salient issues. How do people who do not have access to affordable internet connectivity to public e-services participate digitally? How do digital literacy dynamics play a part in this socio-technological process? What digital literacy competencies and conditions are needed for full participation in a digital society? Is the power of community and community networks one of the solutions? These are some of the questions that this book seeks to answer and highlight. Accordingly, the chapters in the book aim to counteract this with stories on initiatives, platforms, and best practices to help scholars, practitioners, and policymakers. Their propositions and findings provide theoretically grounded and evidence-based research that informs interventions to ensure that all citizens have and can enhance their digital literacy to be able meaningfully and responsibly participate in the digital economy and society.

The challenges lie in the financing, networking, and regulatory environments, as well as deploying and scaling business models that can sustainably provide connectivity and digital literacy initiatives to low-income regions. We hope to see more inclusive solutions, converging practices, and implementing programs on digital literacy skills that are strategically tailored and based on local needs. One of the solutions could be the implementation of digital portals on the national level, with open-source content and digital skills training programs. This could address four critical areas: *access, skills, regulations, and inclusion,* and build the framework for societal empowerment in developing countries. We must focus on higher digital skills standards and policies and a common framework among governments, private companies, civil society, and communities. Moreover, we need strategic efforts that bring to the fore solution to uplift and empower vulnerable and marginalized groups, including children, women, elders, and people with disability, among others. This includes the necessary investments in both human capital and infrastructure, and smart regulatory and standardized environments. Affordable and quality infrastructure, platforms, and digital literacy programs, combined with regulatory frameworks and policy, would enable governments and organizations to participate in the digital economy, helping countries and individuals in their overall well-being and collaboration.

And as the world begins to recover from the widespread effects of global disruption in the past 3 years and at the same time, it is facing the worse humanitarian crisis and geo-political entropy, there is no better timing than this to (re)examine the necessity and impact of digital transformation. Therefore, the book nudges us to ask new questions: What does it take for individuals and communities, and what is required for governments to adapt to the new digital landscape and practices that anticipate the fifth industrial revolution? Could meaningful and affordable connectivity and digital literacy initiatives offer us the solution and inspiration for our renewal, digital resilience, and regeneration? What if renewal and rebirth were encouraged and required for our (digital) society's future?

References

Carretero S, Vuorikari R, Punie Y (2016) Dig- comp 2.1: the digital competence framework for citizens with eight proficiency levels and examples of use. European Commission, Brussels. Retrieved from https://publications.jrc.ec.europa.eu/repository/bitstream/JRC106281/web-digcomp2.1pdf_(online).pdf

Cowley R, Joss S, Dayot Y (2018) The smart city and its publics: insights from across six UK cities. Urban Res Pract 11(1):53–77

Eslami M, Karahalios K, Sandvig C, Vaccaro K, Rickman A, Hamilton K, Kirlik A (2016) First, I "like" it, then I hide it: folk theories of social feeds. In: Proceedings of the 2016 cHI conference on human factors in computing systems, pp 2371–2382

Haythornthwaite C, Andrews R (2011) E-learning theory and practice. SAGE, London

Helsper EJ (2021) The digital disconnect: the social causes and consequences of digital inequalities. SAGE, London

International Telecommunication Union (2022) Measuring digital development: facts and figures 2022. Accessed at https://www.itu.int/en/ITU-D/Statistics/Pages/facts/default.aspx

International Telecommunication Union (2021) Facts and figures 2021. https://www.itu.int/itu-d/reports/statistics/facts-figures-2021/

Karizat N, Delmonaco D, Eslami M, Andalibi N (2021) Algorithmic folk theories and identity: how TikTok users co-produce knowledge of identity and engage in algorithmic resistance. Proceedings of the ACM on Human-Computer Interaction 5(CSCW2):1–44

Kleine D (2013) Technologies of Choice? ICTs, development, and the capabilities approach. MIT Press, Cambridge, MA

Kunze J (2020) Data literacy in the Smart City. Geoforum Perspektiv 19(35). https://doi.org/10.5278/ojs.perspektiv.v19i35.6423

Muringani J, Noll J (2021) Societal security and trust in digital societies: a socio-technical perspective. In: 2021 14th CMI international conference-critical ICT infrastructures and platforms (CMI). IEEE, pp 1–7

Radovanović et al (2015) Overcoming digital divides in higher education: digital literacy beyond Facebook. In: New media & society. SAGE. Available at: http://nms.sagepub.com/content/early/2015/06/02/1461444815588323.abstract

Radovanovic et al (2020) Digital literacy key performance indicators for sustainable development. Social Inclusion 8(2):151–167. https://doi.org/10.17645/si.v8i2.2587

Ragnedda M (2017) The third digital divide. A Weberian approach to digital inequalities, Routledge. Routledge Advances in Sociology

Thomas J, Barraket J, Wilson CK, Holcombe-James I, Kennedy J, Rennie E, Ewing S, MacDonald T (2020) Measuring Australia's digital divide: the Australian Digital Inclusion Index 2020. RMIT and Swinburne University, Melbourne for Telstra

van der Velden M (2018) Digitalisation and the UN sustainable development goals: what role for design. ID&A Interaction Design & Architecture(s) 37:160–174

Van Der Velden M (2021) 'Fixing the world one thing at a time': community repair and a sustainable circular economy. J Clean Prod ISSN 0959-6526:304. https://doi.org/10.1016/j.jclepro.2021.127151

Part I
Digital Literacy Theory Implications

Chapter 2
From Skilled Users to Critical Citizens? Imagining and Future-Making as Part of Digital Citizenship

Johanna Ylipulli and Minna Vigren

Introduction

The profound digitalization of everyday environments and the deep social, political, and economic implications of this development mean there is an urgent need for not just advancing but *rethinking* digital citizenship, as well as reconsidering different actors' roles within this new order. In many societies, digital literacy already determines individuals' possibilities to effectively function as a part of society: everyday life actions such as accessing health services and using public transport are dependent on digital literacy. However, this is just one level of the ongoing transformation. More broadly, digital technology is deeply woven into everyday practices, social interactions, cultural experiences, economic transactions, and political decision-making (Manovich 2013; Williamson 2014). This results in that digital literacy, understood mainly as a set of adequate skills that help in navigating digitalized everyday environments, needs to be accompanied by notions of broader critical awareness of technologies' role in society. In this chapter, we explore one facet of this critical awareness: how the notion of *digital citizenship* could be complemented and expanded to include an *ability to imagine alternative future trajectories*.

The chapter is, above all, a theoretical essay. However, we also provide examples from our previous or ongoing empirical studies to support our arguments. The structure of the text is as follows: we first explain how the depth and breadth of ongoing digital changes motivate us to move, on a conceptual level, from *digital divide to digital inequality*, and from the *digital literacy to digital citizenship*. However, we wish to emphasize that the intention is not to argue for replacing any concept with another, but rather to refocus the discussions in order to provide an alternative,

J. Ylipulli (✉) · M. Vigren
Aalto University, Espoo, Finland
e-mail: johanna.ylipulli@aalto.fi; minna.vigren@aalto.fi

D. Radovanović (ed.), *Digital Literacy and Inclusion*,
https://doi.org/10.1007/978-3-031-30808-6_2

broader framing for the digital literacy of the twenty-first century. Second, we explain how our respective empirical studies focusing on experiences and perspectives of individuals living technologically saturated lives, have informed our understanding, and given insights on how people perceive their technological agency—or the lack of it. We continue by drawing from approaches provided by *design-oriented thinking*, especially *speculative design* and *Participatory Design (PD)* to broaden the discussions linked to digital literacy and digital citizenship towards understanding *active, participatory future-making as a means to increase technological agency and technological awareness. Finally, we provide examples from our recent empirical research to briefly introduce the challenges and possibilities of the approaches drawn from design and focusing on participatory* future-making.

From Digital Divide to Digital Inequality, and from Digital Literacy to Digital Citizenship

There exists a wealth of literature tracing how differing access to and skills in using digital technologies put people in unequal positions in society. The term *digital divide* was coined two decades ago; originally, it was a binary classification pointing to those who have access to the Internet and/or ICTs (information and communication technologies) and those who do not (e.g. Novak and Hoffmann 2000; Stiakakis et al. 2009). After the turn of the century, the term has been extended to somewhat different directions: scholars began referring to the first, second, and third level digital divide: The first level refers usually to access or connectivity, the second level includes digital literacy and skills, and the third level of digital divide describes the broader everyday life benefits and opportunities enabled by internet access and digital skills (Radovanović 2021).

Alongside the term digital divide, the use of the concept of *digital (social) inequality* has increased (e.g. DiMaggio and Hargittai 2001; Durand et al. 2021; Robinson et al. 2015, 2018). Digital inequality highlights, first of all, the ubiquitous nature of digital technology in our society and its numerous entanglements with everyday life; without adequate access, skills, and understanding of the digital realm, it can be difficult to carry out many daily tasks. Second, it effectively emphasizes how the digital realm is creating a new axis of social inequality, which is in many ways linked with the other societal and cultural constructions producing inequalities, such as age, gender, sexual orientation, education, class, and income. Those who are already affected by the conventional forms of inequalities suffer often also from digital inequality, for example, in the form of worse connectivity or weaker skills (Robinson et al. 2015, 2018). These, in turn, can result in many kinds of disadvantages such as not having access to information or education, loneliness, a worse position in labor markets, and difficulties in separating disinformation from more reliable information—just to mention a few examples.

Digital inequality can also be used as a term that reflects wider unequal relationships in society—not just between individuals but also between individuals and large technology companies or (authoritarian) governments. This shifts the focus from the social level to the societal level. The nearly ubiquitous use of digital technologies has given birth to the *data economy* in which free, often quite addictive digital services are provided to billions of people. Concurrently, the data produced through the use of those services is collected, analyzed with algorithms or AI, and sold. The users of the services are provided with explanations about the data collection and use—following also some recent legal measures, such as the *EU General Data Protection Regulation*—but they are often vague and complex, and thus, incomprehensible for most people (Lehtiniemi and Haapoja 2020; Ylipulli and Luusua 2019). Shoshana Zuboff (2015, 2019) has famously coined the term *surveillance capitalism*, which refers to a system in which people's behavior is monitored in astonishing detail by big technology corporations through digital service use, and this surveillance is monetized. To address the presented development and to re-conceptualize the idea of divides in the digital society, Mark Andrejevic has introduced the notion of *Big Data Divide*. By this, he refers to the "asymmetric relationship between those who collect, store, and mine large quantities of data, and those whom the data collection targets" (Andrejevic 2014, p. 1673). The growing role of digital data has profound consequences for democracy: in this new order, data equals knowledge, and knowledge equals power, leading to a deep power asymmetry (Hintz et al. 2017). The ones holding knowledge can have control over the monitored ones, and this manipulation can be carried out in very invisible and subtle ways, by using social media bubbles, targeted news, and targeted disinformation. Public institutions can also contribute to this distortion of power relationships, either inadvertently or on purpose: It has been demonstrated that policy algorithms and predictive risk models can be very biased and discriminatory (Eubanks 2018). Deconstructing the described power asymmetry is by no means easy; for example, Andrejevic (2014) argues that the practices of data mining are dependent on storing infrastructures and analytics skills in a manner that granting access to data for ordinary citizens would not solve the problem of the Big Data Divide.

The definition of digital literacy and its relation to neighboring concepts such as digital citizenship remain nebulous, and the definitions are overlapping, competing, and often divergent (see e.g., Helsper and Eynon 2013; Nichols and Stornaiuolo 2019). Digital literacy has been largely adopted as a target in policy and in educational contexts. In one of its narrowest definitions, digital literacy is considered synonymous with the ability of individuals to participate in the economy through skills and creativity enabled by digital technologies (Klecun 2008; Littlejohn et al. 2012; digital skills in relation to multimodalities, see Radovanović et al. 2020). Such skill-oriented understanding of digital literacy is portrayed for example in the European Union's newest strategic plan titled *2030 Digital Compass: The European Way for the Digital Decade* (European Commission 2021). In this policy document, there is a call for "digitally empowered and capable citizens", but the meaning is reduced to imply a digitally skilled workforce, especially for the IT sector, consumers who trust digital products and online services, and skilled users who identify disinformation, protect themselves from cyber threats, and know how to navigate in online environments. In other words, digital skills are framed as an obligation the citizens must fulfill in order to support economic growth—and this does not include critical thinking nor the capability to evaluate the conditions of the digital society (see also Fraser 2007; Saariketo 2013, 2015).

Although digital literacy has been defined also in broader terms to address questions of representation (whose voices are heard and not heard), language (how digital media is constructed), production (who is communicating to whom and why), and audience (how content is targeted to audiences and how people as audiences use digital media) (Buckingham 2006), the metaphor of literacy is often used as a vague synonym for competence or a skill, with a focus on mastery and operational proficiency, or evaluation and critique (Lankshear and Knobel 2011). The focus on competencies has done well in highlighting a skill set including searching and interpreting data, communicating and collaborating with others, dealing with the negative aspects of online life as well as participating effectively and meaningfully in online activities (see also, Choi and Cristol 2021; Radovanović et al. 2020). However, the emphasis on acquiring skills has consequences for what is considered the ideal of a "digitally literate" person. The competence-oriented understanding tends to idealize production skills; at the same time, it ignores the importance of understanding the social, economic, and cultural context of use of the digital technologies and does not pay sufficient attention to the need for critical thinking skills. To address the complexities of a contemporary digital environment, new literacies have been introduced, such as code literacy (Rushkoff 2010; Vee 2017), algorithmic literacy (Cotter and Reisdorf 2020; Dogruel et al. 2021), data literacy (Livingstone et al. 2020; Pangrazio and Sefton-Green 2020), and data infrastructure literacy (Gray et al. 2018). The prominent problem with these new (and endless) literacies is that they comply with the changes in the media technological environment, and thus imply an orientation towards present-day progress and innovation of technologies. The new literacies tend to prioritize users, devices, and content and simultaneously ignore questions and concerns over the co-constructedness of technology and society, as well as the technical infrastructures and socio-economic relations of living in thoroughly technologized environments (Erstad 2010; Nichols and Stornaiuolo 2019; Njenga 2018). This has meant that the definitions of digital literacy have often lacked contextualization as well as a focus on the ideology and power manifested in the design and use of technologies. Consequently, they have also bypassed the situatedness of digital realities by ignoring the experiences of the marginalized and impoverished people in different parts of the world (Choi and Cristol 2021), and the dilemmas connected to surveillance, control, and datafication in the networked environments. We suggest that (at least some of) these shortcomings could be tackled by drawing ideas and approaches from design studies and by highlighting the role of active imagining and alternative futures, as explained in more detail in the next section. In other words, we intend to expand the meaning of the term digital literacy. However, we also wish to underline the usefulness of the concept of digital citizenship.

Whereas some consider the concepts of digital literacy and digital citizenship almost synonymous (e.g. Knox and Bayne 2013), others see digital literacy as foundational for digital citizenship: literacy and implied skills of reading and writing about the digital are considered as a pre-condition for participating in the digital society and claiming digital rights (Pangrazio and Sefton-Green 2021). The terminology relying on the classical view on citizenship has been justly criticized due to its narrow and biased approach to identity and participation: it has ignored the varying forms of social inequalities, as well as their interrelation to intersectional attributes (e.g. ethnicity, gender, class, age, and sexuality), the role of situated struggles, and experiences of marginalization, oppression, and vulnerability (see also Choi and Cristol 2021). Despite these criticisms, we see benefits in approaching agency in the digital society with the notion of digital citizenship as it connects the considerations of individual micro-level agency and societal macro-level actions. We embark from the definition by Hintz and colleagues according to which digital citizenship is "*based on the possibility of comprehensive self-determination in a datafied environment, provided by secure infrastructure, an enabling regulatory environment, adequate public knowledge, and an informed use of the relevant platforms and applications*" (Hintz et al. 2019, p. 41). This definition differs from the narrower ones in the sense that it underlines the accountability of society and its various institutions. Becoming capable of making sense of the digital everyday life and acting accordingly does not just depend on the individual but society as a whole has responsibilities concerning digitalization—it is obligated to protect democracy and resist the tendencies that are strengthening social and societal inequalities. Carefully crafted ethical guidelines, new laws, and regulations are needed not only to guarantee secure infrastructure, but also sustainable, fair, and transparent infrastructures. Thus, the concept of digital citizenship captures the ontological definition of citizenship which marks the relationship to something beyond the individual—be that the community, platform, or the nation-state (McCosker et al. 2016); it emphasizes that digital citizenship is not just about civic responsibilities or self-responsibilization.

Towards Stronger Agency Instead of Enhanced Access and Skills?

In this section, we tie together some observations and reflections from our previous studies through the concept of *critical agency* by Paola Rebughini (2018). The notion of critical agency offers a fruitful conceptual basis for elaborating a more holistic approach to digital citizenship as it underlines critical thinking and reflection as affirmative creation of new practices. Rebughini's definition of the concept goes beyond understanding agency solely as the capacity of a subject to act in an autonomous way. She brings forth the ideas of dissident, innovative, imaginative, and transformative aspects of agencies that orient "against and beyond what is perceived as unjust, unequal, [and] unacceptable" (Rebughini 2018, p. 3). We embrace Rebughini's idea of critical agency as a capacity to innovate and produce new practices and imaginaries of alternative futures that bypass (and not only unmask) constraints and mechanisms of contemporary (digital) society. Further, we introduce approaches from the design studies that can promote the construction of critical agency through participatory actions.

Our first empirically embedded notions on agency are from a qualitative study carried out by the first author of this article roughly 10 years ago, in 2011–2012, in which young adult study participants (n = 48, aged 20–30 years) were studied in order to find out (1) how they use digital technologies, especially ICTs, in their everyday lives, (2) how they experience their current use and the role of technology, and (3) what kind of dreams and fears they have concerning the future technology (Ylipulli 2015a, b). The results pertaining to the last part indicate that the need to focus on enhancing people's agency over technologies is not just a creation of experts but the same is expressed also by "lay" people, living technologically saturated lives. A significant part of the studied young adults described their dream technology with expressions and imagery connected to *nature, naturalness, calmness,* and *unobtrusiveness*. However, they did not wish that technology would be an invisible helper that is automatically performing tasks for them, like in many prevailing technology visions, such as *ubiquitous computing* or *context-aware computing* (Reeves 2012). Rather, they hoped to stay in control and use technology in a controlled manner. Some participants also directly stressed that there is a need for *critical reflection*, and stated that people should realize the role technology actually plays in their lives. They were argumenting that only this type of awareness would, in turn, enable them to "fuse technology in their lives calmly" (female participant, aged 26). A more detailed analysis is presented in Ylipulli (2015a), and to summarize, the studied young adults clearly sought to *keep their agency* and not transfer it to computers. Interestingly, also a need to better understand the various impacts of technology on everyday life surfaced in the study repeatedly.

In her dissertation, the second author studied how the imaginaries of one's agency and the agency of others are constructed and stabilized in thoroughly digitalized everyday life (Saariketo 2020). The question was tightly related to the question of contemporary societal power structures and arrangements, and how their production, reproduction, and contestation intertwine with the processes of constructing imaginaries of agency. The findings indicated that many of the 33 research participants experienced deep feelings of helplessness, frustration, and a resigned sense of agency in relation to their digital everyday life (anonymized; see also Andrejevic 2014; Markham 2021). This resigned sense of agency is supported, and partly constructed, in the interpellations to the agency by the administration-political discourse and mediated pre-domestication of new technology. In these dominant discourses, citizens are persuaded to adopt a form of agency that promotes increased consumption, economic growth, and simultaneously reduces criticism of the (infrastructural) conditions of digital everyday life. These rather narrow roles of agency offered to people introduce technological development and the embedded values of technology as taken-for-granted, contributing to the condition in which questions of the costs of connectivity and its infrastructural nature are pushed to the background and disappear from sight. Despite the sporadic negotiations and dissonances that surface in the empirical research data, it seems that many people have grown accustomed to the notion that they have very little or no opportunity to influence the structural conditions of their digital environments. Thus, visions that could challenge or radically alter the sociotechnical forces that currently condition agency remain in the margins. This underlines an urgent need to imagine alternatives and support people's sense of agency in a manner that lets people imagine what kind of technology they want to live with and create space to act alongside these visions.

We argue the above-presented observations from both studies point towards a focal point of digital citizenship: we must find ways to enhance people's agency by fostering competencies related to *imagining alternative futures*. Our own understanding of this approach arises from the field of design, and we perceive it as an awareness of the contingent nature of the future, and as practices of using imagination and creativity consciously in different processes and activities related to the future, with the help of various techniques and tools. Imagining alternative futures is closely related to *futures thinking,* a term often used in Future Studies and strategy work, which has also recently found its way into pedagogical contexts (e.g. Häggström and Schmidt 2021; Levrini et al. 2019; Rasa et al. 2022). In the following, we introduce two approaches from the field of design that offer beneficial perspectives for active and conscious future-making: *Participatory Design* (PD) and *speculative design*.

In design research focused on new technologies and in related fields, such as human-computer interaction (HCI), there is a wealth of literature concerning technology users' empowerment and strengthening of agency. The Scandinavian tradition of Participatory Design is perhaps most explicitly anchored in empowering technology users (e.g. Björgvinsson et al. 2010, 2012). The aim of a PD process is usually to create a technological concept or an artifact through multi-stakeholder collaboration in which the so-called end-users have a significant role. Originally, when computer systems were designed mainly for work environments, the intention was to empower workers. Since those early years, computers have spread to almost all areas of life, and thus also PD is nowadays practiced within different contexts and with different kinds of groups of people. Blomberg and Karasti (2012) list the central principles of PD as follows: The participants must respect different kinds of knowledge; the process must offer opportunities for mutual learning, joint negotiation of project goals, as well as tools and processes to facilitate design. It is important to note that the goal of the design process is not only the designed product: *It is also of crucial importance that through processes of mutual learning participants gain insights into design processes, begin to understand the impacts of technology, and realize they have a choice* (Ylipulli et al. 2017). These principles are already combined with educational aims through work that focuses on fabrication laboratories (fablabs) and enhancing children's making and design skills in school contexts (e.g. Iivari and Kinnula 2018).

In addition to PD, we find the perspective offered by *speculative design* important. Speculative design is one of the somewhat rebellious branches of design research, part of a broader discourse of *critical design* born in the 1990s. Anne Galloway (2013) comments that criticality in critical design is framed in a way that makes it actually feel familiar to social scientists: design work is understood as a cultural commentary, not necessarily as a functional prototype. The intention is to ask questions rather than give answers. During recent years, approaches related to speculative design have gained traction. Anthony Dunne and Fiona Raby (2013), the well-known proponents of speculative design, state that *"This form of design thrives on imagination and aims to open up new perspectives on what is sometimes called wicked problems, to create spaces for discussion and debate about alternative ways of being, and to inspire and encourage people's imaginations to flow freely. Design speculations can act as a catalyst for collectively redefining our relationship with reality."* Speculative design comes in many forms: it can be an object, a website, a fictional narrative, or a process that does not result in an artifact of any kind (Blythe 2014; Earth 2050; Ylipulli et al. 2016). What is common for all these approaches is to address the current reality by imagining alternatives. The speculative design has been criticized for being elitist and confined to galleries and universities (e.g. Mazé 2016). However, the more recent trends in the field argue for combining *participatory approaches to design* with speculative approaches which enables imagining new futures together with a more diverse group of participants (Baumann et al. 2018; Lyckvi et al. 2018; Rüller et al. 2022).

These approaches, and especially their intersections, can provide fresh viewpoints for understanding criticality and active imagining of alternative future trajectories as part of digital citizenship. They are echoing the notions presented by Rebughini (2018); these designerly approaches can foster imaginative and transformative aspects of the agency. They are also highly flexible and can work in different kinds of contexts and with different groups of people. In addition, design is a very practice-oriented field, and thus, it offers us numerous detailed, practical methods to be utilized and applied in activities, events, and education connected to emerging technologies and digitalizing society.

Strengthening Technological Agency by Creating Opportunities for Participatory Imagining of Alternative Futures

Understanding active imagining and future-making as a part of digital citizenship is exemplified below through more recent empirical studies we have conducted, and in which we have drawn from PD and speculative design. The studies have been carried out in different contexts and with different collaborators, underlining the flexibility of the suggested approaches and also highlighting our intention to democratize future-making activities. Our scope is not limited to conventional educational contexts but we stress that people of all ages should have the possibility to become "digital citizens". This means that there should be an emphasis on lifelong learning possibilities, but at the same time, it points towards structural and paradigmatic changes not just in formal education, but also in how we design technologies and what kind of roles and responsibilities are assigned to powerful institutional and commercial players of digitalization.

The first empirical study we briefly present is actually not a singular study, but a long-term collaboration process unofficially titled *the Virtual Reality Library*, carried out jointly by a network of public libraries from five different cities and two different universities in Finland. The process started already in 2016 when Oulu City Library and the University of Oulu began to collaborate in Northern Finland and has since expanded to cover all the libraries in the capital region as well as Aalto University. The experimentation with VR has received funding from several different sources, and at the time of writing this chapter, it is still ongoing. The concrete aim of the process has been to produce a functional Virtual Reality (VR) application for the use of libraries. In Finland, public libraries are relatively well-funded and respected institutions (Vakkari and Serola 2012), and they are given a rather central educational role in society: the current legislation, the *Public Libraries Act*, defines libraries as sites that should provide people means to participate in different societal discussions as active citizens. Education is understood as an enabler of active citizenship and it is also seen as a prerequisite for all action. This educational objective also includes an obligation to educate citizens about new technologies. Currently, especially the large libraries of the metropolitan region are well equipped for this mission having an impressive infrastructure as well as a variety of events and courses related to new technology (Ylipulli and Luusua 2019).

The *Virtual Reality Library* process is based on applying Participatory Design and speculative design in a real-world context, resulting in a functional artifact but also a mutual learning process. The design and development have followed the principles of PD in the sense that library staff members have been participating continuously from the beginning: They have been shaping the aims of the project as well as defining the visual appearance and functionalities of the application. Furthermore, library patrons of different ages have been part of the design process through five multi-stakeholder workshops and two sets of user tests conducted during the research and design work (Pouke et al. 2018; Ylipulli et al. 2020). Speculative design and imagining alternative futures were used in the workshops where we intended to foster participants' creativity by using fiction as a source and also so-called *creative metaphors* as a method (Ylipulli et al. 2017). At the time of writing this Chapter, the process has resulted in a multi-purpose VR application that can be used with Oculus Quest headset and controllers, and which is offered for all the public libraries in the country for free. The application consists of three different virtual environments presenting different forests where the user can just move around or play a game consisting of different tasks, such as archery, and collect and arrange literary quotes (Figs. 2.1 and 2.2).

One of the main aims behind creating a VR application tailored for public libraries is to provide them with a tool for media and technology education. Using PD combined with speculative design as an approach resulted in an application that libraries actually find usable, which is aligned with their mission and societal tasks,

Figs. 2.1 and 2.2 In the Virtual Reality Library, the user can wander through environments representing different kinds of forests, from a cartoonish winter forest to a more realistic one resembling a Finnish forest. The environments are based on the design participants' ideas. (Images: Center for Ubiquitous Computing, University of Oulu)

and which is based on creative ideas created by library staff and library patrons themselves (Fig. 2.3). The process itself offered opportunities for learning about (1) VR technology, (2) the technology design process, and (3) future-oriented (design) thinking. Furthermore, the participants of the process were able to experience how their abstract and even wild imaginings were turned into a concrete technological application from scratch. Of course, these possibilities that can potentially orient the participants towards a more active future-making and strengthen their agency were available only for a limited number of people, as the workshops and user test events needed to be kept relatively small. However, the libraries plan to use the application

Fig. 2.3 The VR application in question is tailored for public libraries and it provides them a tool for media and technology education. The content is connected to the library context; the user can complete gamified tasks and collect literary quotes from a magical book. The quote is from the Finnish translation of Jill Barklem's children's book *Autumn Story* (1980). (Image: Center for Ubiquitous Computing, University of Oulu)

with different kinds of groups (young people with refugee backgrounds, the elderlies) to enhance their understanding of the new media and technology. It can be used and played as it is, or it can be complemented with new virtual objects or complete VR environments as the code is open. Thus, the libraries can continue designing the application and related practices further.

The process is an example of a participatory, speculative future-making activity that is also scalable. We are currently conducting post-design process interviews with the main stakeholders, but we have not specifically studied whether the participants experienced their participation as empowering and did it transform their ideas about the future. However, this is indicated in the literature (Hansen et al. 2019). Further, we do not know the broader impact of the process yet, as the libraries are currently appropriating the VR application.

Our second case study is a project called *Young people imagining alternative media(ted) futures,* which focused on studying what kind of mediated everyday life young people would like to have. The starting point was the observation that it is very difficult to take critical distance from and to envision alternatives to, the ways things "normally" are in the thoroughly networked everyday life (Saariketo 2020; Markham 2021). The study focused on young people as it was contended that the taken-for-grantedness is particularly pronounced within an age group that has grown surrounded by the ubiquity of digital technologies. It was also acknowledged that the future concerns young people in a particular manner, but they are usually not heard in discussions about the future.

During the one-year project (2020–2021), an *Imagining Workshop* model titled *Media/The Everyday Life 2030* was developed. To facilitate designing the workshop, we conducted an online survey ($n = 436$) with young people to serve as background material. The aim of the workshop was to enable the young participants aged 14–18 ($n = 24$) to distance themselves from the self-evident aspects of their mediated everyday life and encourage them to imagine jointly alternative mediated futures. In the online workshop, the participants worked in small teams, and with the help of different methods such as character creation and shuffled cards, the participants imagined little scenes happening in the future, in 2030. The cards helped them to imagine changes in the media environment and related moods.

The young participants felt that they were heard during the workshops. The playing cards gave space for them to express their concerns, such as environmental issues, dataveillance, fake news, oppression, and new forms of slavery, as well as online and offline harassment and hate speech. One of the biggest concerns was a fear that some of the main problems in 2021 would still be issues in 2030. A major observation from the workshops, and one that fosters hope for the future, concerns the flexibility of imagining as a joint activity. While collectively projecting alternatives to the mediated everyday life proved challenging for the participants, it sparked vibrant discussions on the bleaker aspects of the contemporary networked society. However, future empirical research needs to explore further the critical potential inherent in the collective exercise of human imaginative capacities. One challenge relating to this is that the act of imagining is rooted in the lived experiences of the past and does not necessarily result in the expression or co-production of counter-hegemonic narratives of the future. Imagination as such does not necessarily include a transformative aspect, and focus should be laid on *aspiration*—the ideas of how the future should differ from the present day (see also Appadurai 2013).

Discussion and Conclusions

To conclude, in this chaper, we have intended to demonstrate how the ubiquitous role of digital technologies in contemporary society calls for novel understandings of digital literacy. Digital technology contributes to societal inequality in numerous ways, creating divides not just between different groups of people but also between people and institutions, such as large companies and governments. Instead of focusing on individuals' digital skills and their ability to adapt to the prevailing societal reality, defined strongly by digitalization, we have framed the digital literacy of the twenty-first century as digital citizenship and explored the possibility to understand imagining and future-making as important facets of it. Digital citizenship as we understand it draws from the broad definition presented by Hintz et al. (2019) which grants the individual a "*comprehensive self-determination in a datafied environment*" but highlights also the role of society in creating suitable conditions (Hintz et al. 2019, p. 41). This definition does not cast all the responsibility of becoming a 'proper digital citizen' on the individual. In other worlds, it is imperative that society provides opportunities for learning digital skills for people of all ages, but we argue that there needs also to be possibilities to learn critical thinking skills: how to question the prevailing developments—and how to imagine alternatives.

Although researchers in the field of digital literacy have acknowledged "*the need to imagine multiple futures*" (Njenga 2018, p. 4) and the potential in intervening "*the systems that produce them [new technologies] in order to make them more just and equitable*" (Nichols and Stornaiuolo 2019, p. 21), it has seldom been explored what this means in practice. We have drawn from the field of design studies and introduced especially Participatory Design and speculative design as potential approaches for incorporating active, participatory future-making as part of digital citizenship. These approaches can be utilized in various contexts and with various groups of people, as our two case study examples demonstrate: the first focuses on the context of public libraries, covering thus library staff and (potentially) all the library patrons, and the second was centering on adolescents and their ability to imagine alternative digital futures.

Our proposed understanding of digital citizenship goes beyond teaching people how to use and appropriate existing technologies in the digital society. Thus, it goes beyond contributing to the reproduction and stabilization of digital technology and the implied power arrangements as part of daily life. We understand digital citizenship as a fundamentally political practice that acknowledges the foundations and implications of the development and application of digital technology in our lives. There is a need to reflect upon, challenge, and resist the kind of oblivion that prevents us from seeing how things could be otherwise. With the means of active future-making approaches, we have introduced an idea of digital citizenship that opens the sociotechnical construction of digital for negotiation. It is obvious that the ideas and alternatives we have briefly introduced need to be conceptualized further, and also their potential must be studied through empirical research. One of the issues that need to be addressed in the future is the sheer complexity and black-boxed nature of digital environments which poses severe challenges (even for experts!) to understanding and knowing how digital technology works. After all, knowledge and understanding have been considered as important building blocks of citizenship in a technological society (Feenberg 2011; Isin and Ruppert 2017).

We can conclude here that the capability to reflect upon the foundations of (digital) societies does not depend on technical expertise: even without possessing detailed technological knowledge, people are capable of reflecting on what kind of society they would like to live in. We believe that curious and fearless imagination can enable awareness, reflection, challenging of, and resistance to the conditions of digital environments; it can lead us beyond the taken-for-grantedness, and thus give space for visions of alternative digital futures.

Finally, we wish to highlight that digital citizenship must include a critical approach to both discursive and material construction of technology and the complex web of exploitation linked with digital technologies (see also Emejulu and McGregor 2019). This means awareness of the material underpinnings of digitality, including the natural resources and labor needed in producing the devices and recycling the e-waste that all have political and environmental consequences in the Global North, and especially in the Global South.

Acknowledgements This work was supported by the Academy of Finland funded projects Digital Inequality in Smart Cities, DISC, (332143), and Imagining Sustainable Digital Futures (347950).

References

Andrejevic M (2014) The big data divide. Int J Commun 8:1673–1689
Appadurai A (2013) The future as cultural fact: essays on the global condition. Rass Ital Sociol 14(4):649–650
Baumann K, Caldwell B, Bar F, Stokes B (2018) Participatory design fiction: community storytelling for speculative urban technologies. In: Extended abstracts of the 2018 CHI conference on human factors in computing systems. ACM Press, New York, pp 1–1
Björgvinsson E, Ehn P, Hillgren PA (2010) Participatory design and "democratizing innovation". In: Proceedings of the 11th biennial participatory design conference. ACM Press, New York, pp 41–50
Björgvinsson E, Ehn P, Hillgren PA (2012) Agonistic participatory design: working with marginalised social movements. CoDesign 8(2–3):127–144
Blomberg J, Karasti H (2012) Positioning ethnography within participatory design. In: Simonsen J, Robertson T (eds) Routledge handbook of participatory design. Routledge, New York, pp 86–116
Blythe M (2014) Research through design fiction: narrative in real and imaginary abstracts. In: Proceedings of the SIGCHI conference on human factors in computing systems. ACM Press, New York, pp 703–712
Buckingham D (2006) Defining digital literacy: what do young people need to know about digital media? Nordic J Digit Lit 1(4):263–277
Choi M, Cristol D (2021) Digital citizenship with intersectionality lens: towards participatory democracy driven digital citizenship education. Theory Pract 60(4):361–370
Cotter K, Reisdorf BC (2020) Algorithmic knowledge gaps: a new horizon of (digital) inequality. Int J Commun 14:745–765
DiMaggio P, Hargittai E (2001) From the 'digital divide' to 'digital inequality': studying Internet use as penetration increases, vol 4(1). Center for Arts and Cultural Policy Studies, Woodrow Wilson School, Princeton University, Princeton, pp 4–2
Dogruel L, Masur P, Joeckel S (2021) Development and validation of an algorithm literacy scale for internet users. Commun Methods Meas 16(2):115–133
Dunne A, Raby F (2013) Speculative everything: design, fiction, and social dreaming. MIT Press, Cambridge, Massachusetts
Durand A, Zijlstra T, van Oort N, Hoogendoorn-Lanser S, Hoogendoorn S (2021) Access denied? Digital inequality in transport services. Transp Rev 42(1):32–57
Earth 2050 by Kaspersky. Accessed 17 Feb 2022. Retrieved from https://2050.earth/
Emejulu A, McGregor C (2019) Towards a radical digital citizenship in digital education. Crit. Stud. Educ 60(1):131–147
Erstad O (2010) Educating the digital generation. Nordic J Digit Lit 5(1):56–72
Eubanks V (2018) Automating inequality: how high-tech tools profile, police, and punish the poor. St. Martin's Press, New York

European Commission (2021) 2030 Digital Compass: the European way for the Digital Decade. Communication from the Commission to the European Parliament, the Council, The European Economic and Social Committee and the Committee of the Regions. Retrieved from https://eur-lex.europa.eu/legal-content/en/TXT/?uri=CELEX%3A52021DC0118
Feenberg A (2011). Agency and citizenship in a technological society. Lecture at the IT University of Copenhagen. Retrieved from https://www.sfu.ca/~andrewf/copen5-1.pdf
Fraser N (2007) Creating model citizens for the information age: Canadian internet policy as civilizing discourse. Can J Commun 32:201–218
Galloway A (2013) Emergent media technologies, speculation, expectation and human/nonhuman relations. J Broadcast Electron Media 57(1):53–65
Gray J, Gerlitz C, Bounegru L (2018) Data infrastructure literacy. Big Data & Society, July–December
Häggström M, Schmidt C (2021) Futures literacy – to belong, participate and act!: an educational perspective. Futures 132:102813
Hansen NB, Dindler C, Halskov K, Iversen OS, Bossen C, Basballe DA, Schouten B (2019) How participatory design works: mechanisms and effects. In: Proceedings of the 31st Australian conference on human-computer-interaction, pp 30–41
Helsper EJ, Eynon R (2013) Distinct skill pathways to digital engagement. Eur J Commun 28:696–713
Hintz A, Dencik L, Wahl-Jorgensen K (2017) Digital citizenship and surveillance. Digital citizenship and surveillance society—introduction. Int J Commun 11:9
Hintz A, Dencik L, Wahl-Jorgensen K (2019) Digital citizenship in a datafied society. Polity Press, Cambridge, UK
Iivari N, Kinnula M (2018) Empowering children through design and making: towards protagonist role adoption. In: Proceedings of the 15th participatory design conference: full papers-volume 1, pp 1–12
Isin E, Ruppert E (2017) Citizen snowden. Int J Commun 11:843–857
Jorgensen, K. (2017). Digital citizenship and surveillance. Digital citizenship and surveillance society—introduction. International Journal of Communication, 11, 9
Klecun E (2008) Bringing lost sheep into the fold: questioning the discourse of the digital divide. Inf Technol People 21:267–282
Knox J, Bayne S (2013) Multi-modal profusion in the literacies of the Massive Open Online Course (MOOC). Res Learn Technol 21:1–14
Lankshear C, Knobel M (2011) New literacies: everyday practices and social learning, 3rd edn. Open University Press, UK
Lehtiniemi T, Haapoja J (2020) Data agency at stake: MyData activism and alternative frames of equal participation. New Media Soc 22(1):87–104
Levrini O, Tasquier G, Branchetti L, Barelli E (2019) Developing future-scaffolding skills through science education. Int J Sci Educ 41(18):2647–2674
Littlejohn A, Beetham H, McGill L (2012) Learning at the digital frontier: a review of digital literacies in theory and practice. J Comput Assist Learn 28(6):547–556
Livingstone S, Stoilova M, Nandagiri R (2020) Data and privacy literacy. The role of the school in educating children in a datafied society. In: Frau-Meigs D, Kotilainen S, Pathak-Shelat M, Hoechsmann M, Poyntz SR (eds) The handbook of media education research. Wiley, New Jersey, U.S., pp 413–426
Lyckvi S, Roto V, Buie E, Wu Y (2018) The role of design fiction in participatory design processes. In: Proceedings of the 10th Nordic conference on human-computer interaction, pp 976–979
Manovich L (2013) Software takes command. Bloomsbury, London
Markham A (2021) The limits of the imaginary. New Media Soc 23(2):382–405
Mazé R (2016) Design and the future: temporal politics of 'making a difference'. In: Smith RC, Vangkilde KT, Kjærsgaard MG, Otto T, Halse J, Binder T (eds) Design anthropological futures. Bloomsbury, London, pp 37–54

McCosker A, Vivienne S, Johns A (2016) Negotiating digital citizenship: Control, Contest and Culture. Rowman & Littlefield, London
Nichols TP, Stornaiuolo A (2019) Assembling "digital literacies": contingent pasts, possible futures. Media Commun 7(2):14–24
Njenga J (2018) Digital literacy. A quest for an inclusive definition. Read Writ J Read Assoc S Afr 9(1):1–7
Novak TP, Hoffman DL: Bridging the Digital Divide (2000) The internet of race on computer access and internet use. Retrieved from http://www2000.ogsm.vanderbilt.edu/digital.divide.html
Pangrazio L, Sefton-Green J (2020) The social utility of 'data literacy'. Learn Media Technol 45(2):208–220
Pangrazio L, Sefton-Green J (2021) Digital rights, digital citizenship and digital literacy: what's the difference? J New Approaches Educ Res 10(1):15–27
Pouke M, Ylipulli J, Minyaev I, Pakanen M, Alavesa P, Alatalo T, Ojala T (2018) Virtual library: blending mirror and fantasy layers into a vr interface for a public library. In: Proceedings of the 17th international conference on mobile and ubiquitous multimedia. ACM Press, New York, pp 227–231
Radovanović D (2021) In the search of freedom to information access and digital resilience. Springer. Retrieved from: https://www.springernature.com/gp/researchers/the-source/blog/blogposts-life-in-research/freedom-to-information-access-and-digital-resilience/19701156. Accessed 3 Feb 2021
Radovanović D et al (2020) Digital literacy key performance indicators for sustainable development. Soc Incl 8(2):151–167
Rasa T, Palmgren E, Laherto A (2022) Futurising science education: students' experiences from a course on futures thinking and quantum computing. Instr Sci 50(3):425–447
Rebughini P (2018) Critical agency and the future of critique. Curr Sociol 66(1):3–19
Reeves S (2012) Envisioning ubiquitous computing. In: Proceedings of the SIGCHI conference on human factors in computing systems, pp 1573–1582
Robinson L, Cotten SR, Ono H, Quan-Haase A, Mesch G, Chen W et al (2015) Digital inequalities and why they matter. Inf Commun Soc 18(5):569–582
Robinson L, Chen W, Schulz J, Khilnani A (2018) Digital inequality across major life realms. Am Behav Sci 62(9):1159–1166
Rüller S, Aal K, Tolmie P, Hartmann A, Rohde M, Wulf V (2022) Speculative design as a collaborative practice: ameliorating the consequences of illiteracy through digital touch. ACM Trans Comput Hum Interact 29(3):1–58
Rushkoff D (2010) Program or be programmed: ten commands for a digital age. OR Books, New York
Saariketo M (2013) Tulevaisuuden ihannetoimijan tarinallinen tuottaminen Euroopan digitaalisstrategiassa. Hallinnon Tutkimus 32(4):270–283
Saariketo M (2015) Reflections on the question of technology in media literacy education. In: Kotilainen S, Kupiainen R (eds) Reflections on media education futures. Nordicom, Göteborg, Sweden, pp 51–61
Saariketo M (2020) Kuvitelmia toimijuudesta koodin maisemissa. PhD dissertation, Tampere University
Stiakakis E, Kariotellis P, Vlachopoulou M (2009) From the digital divide to digital inequality: a secondary research in the European Union. In: International conference on e-democracy. Springer, New York, pp 43–54
Vakkari P, Serola S (2012) Perceived outcomes of public libraries. Libr Inf Sci Res 34(1):37–44
Vee A (2017) Coding literacy: how computer programming is changing writing. MIT Press, Cambridge, Massachusetts
Williamson B (2014) Governing software: networks, databases and algorithmic power in the digital governance of public education. Learn Media Technol 40(1):83–105
Ylipulli J (2015a) A smart and ubiquitous urban future? Contrasting large-scale agendas and street-level dreams. Observatorio (OBS*) 9:85–110

Ylipulli J (2015b) Smart futures meet northern realities: anthropological perspectives on the design and adoption of urban computing. PhD dissertation, University of Oulu

Ylipulli J, Luusua A (2019) "Without libraries what have we?" Public libraries as nodes for technological empowerment in the era of smart cities, AI and big data. In: Proceedings of the 9th international conference on communities & technologies-transforming communities. ACM Press, New York, pp 92–101

Ylipulli J, Kangasvuo J, Alatalo T, Ojala T (2016) Chasing digital shadows: exploring future hybrid cities through anthropological design fiction. In: Proceedings of NordiCHI '16: Nordic conference on human-computer interaction. ACM Press, New York, Article No. 78

Ylipulli J, Luusua A, Ojala T (2017) On creative metaphors in technology design: case "magic". In: Proceedings of the 8th international conference on communities and technologies. ACM Press, New York, pp 280–289

Ylipulli J, Pouke M, Luusua A, Ojala T (2020) From hybrid spaces to "imagination cities": a speculative approach to virtual reality. In: Willis K, Aurigi A (eds) The Routledge companion of smart cities. Routledge, London, pp 312–331

Zuboff S (2015) Big other: surveillance capitalism and the prospects of an information civilization. J Inf Technol 30(1):75–89

Zuboff S (2019) The age of surveillance capitalism: the fight for the future at the new frontier of power. Profile Books, New York

Chapter 3
Sensing the City: A Creative Data Literacy Perspective

Anne Weibert and Maximilian Krüger

Introduction

As they grow up and seek to find "their" place, children and youth are confronted with several complex phenomena. Where, what, and how to learn and work; digitalization issues; environmental issues; migration and matters of cultural, social, and religious diversity, all these can touch upon every aspect of everyday life. In the urban, children and youth are "designed out" of many physical places (e.g. Hörschelmann and van Blerk 2013). Nonetheless, they are shaping the urban space (e.g. Chawla 2002; Holloway and Valentine 2000). The rural on the other hand, sees children and youth be contained in stable community and family structures (e.g. Panelli et al. 2007). But changing work structures and economies are powerful driving forces, calling for the young to turn their backs on the rural and seek their futures in the urban (e.g. McGrath 2001).

Computing can be a means to assess, express, and ease some of this complexity. Digital skills have long been recognized as a key qualification needed in the modern world (Sefton-Green et al. 2009). Digital and data literacy (DiSessa 2001; Schüller et al. 2019) are each as central a capability in the information society as the ability to read, write, and calculate. Janet Wings work on *Computational Thinking* (2006) marks a central point, bringing forward skills like logically analyzing and organizing data, visualizing data through abstraction, finding efficient solutions, using algorithms to automate solutions, and transferring solutions into other contexts (see also Barr and Stephenson 2011). This perspective also became influential in the learning sciences and educational domain (e.g. Lockwood and Mooney 2017), and the promotion of computational thinking skills in young learners between the ages

A. Weibert (✉) · M. Krüger
Institute for Information Systems and New Media, University of Siegen, Siegen, Germany
e-mail: anne.weibert@uni-siegen.de

D. Radovanović (ed.), *Digital Literacy and Inclusion*,
https://doi.org/10.1007/978-3-031-30808-6_3

of 5 and 18 is now often demanded (e.g. Guzdial 2008). However, traditional methods for teaching computational thinking were found to be not very suitable for children (e.g. Boy 2013), especially for those children with more creative and nonlinear learning types. If computing was truly for everyone, then how could it be ensured that everyone was able to develop the skill set needed to participate? The human senses (Merleau-Ponty 2013) are at the core of discourse lines unfolding from there, touching upon questions of making sense of and participating in (urban) places and communities (e.g. Wolff et al. 2016).

This study argues for the inclusion of making and crafting as alternate methods for urban data literacy—the empowerment of people to solve real-world problems and make sense of what Lohr (2009) has called "the raw material of knowledge" by using and analyzing data from their everyday life in the city, thus measuring and addressing the underlying phenomena in the urban sphere. Our chapter first provides an overview on relevant works from the discourse on human sense-making and (digital) literacy in relation to crafting and making. The description of methods used as well as the setting of the three workshops is followed by our presentation of the findings. Their discussion sheds light on how making and crafting can be a creative means to foster the building of urban data literacy.

Related Works

The importance of materials for human sense-making was recognized in a methodological procedure(s) in various contexts (e.g. Pink 2015; Woodward 2019). Gabrys focused on nature itself, bringing forward environmental as well as sensor-based aspects of ubiquitous computing technologies and related sense-making practices (Gabrys 2016). Her participatory view on matters of "urban sensing" discusses the interrelation of sense-making practices and their computational sensor-based counterpart in the urban sphere. Along similar lines, Mattern has argued that the multifaceted nature of urban data calls for "a degree of sensitivity that exceeds mere computation; urban intelligence of this kind involves site-based experience, participant observation and sensory engagement" (Mattern 2021:70). A number of educational initiatives and tools have taken this perspective to the classroom (e.g. Fauville et al. 2014) and beyond, thus enabling connecting with nature (Rodgers et al. 2019) through technology in material, haptic (e.g. Soro et al. 2018) and programmable ways (e.g. Bröring et al. 2011).

Tangibles were found to be supportive for computational learning and beyond. They allow for the young learners' engagement in problem-solving with concrete physical objects, thus building "representational mappings that serve to underpin later more symbolically mediated activity after practice and the resulting 'explicitation' of sensorimotor representations" (O'Malley and Fraser 2004: 3). "Hands-on" (e.g. Dewey 1923) learning approaches focus on active experimentation with physical materials, often employed in the context of current science education (e.g. Antle et al. 2011). From a constructionist perspective (Papert 1980), knowledge is not only conveyed in the abstract, but requires practical and cognitive (re-)construction by the learner. Learning, in Papert's view, is the process of creating artifacts of personal and social relevance; it is concerned with the connection of old and new knowledge, and with the interaction with others (for an overview see Kafai and Burke 2014: 19ff). Three perspectives are prominent in existing learning and design frameworks. Here, tangibles: (1) make use of physical objects as tokens to access digital information (Holmquist et al. 1999), (2) employ physical objects as containers to move information between devices (Ullmer et al. 2005), and (3) contain tangible interfaces where physical artifacts both represent and control the digital information (e.g. Ullmer and Ishii 2000). Calling for more work on the benefit of physical materials for learning, Marshall provided guidance for the deployment of tangible interfaces for learning (Marshall et al. 2010). Xie et al. (2008) saw the enabling of collaboration among their supportive characteristics, and Hamidi and Bajlko (2017) found their entangling with nature to be supportive for learning. Dourish has described the tangible approach to computing as part of a movement in Human–Computer Interaction seeking to broaden the range of human abilities available when interacting with computers (Dourish 2004).

At the basis of this tangible perspective is touch, which has long been discussed as a cultural technique (e.g. Ufan 1973). Touch can trigger the most basic sensemaking "seeing through the hand" (Hansen 2006: 71), as well as include complex capacity for feelings (McLuhan 1994: 314), in that case not even having an "obvious 'seat' or organ" (O'Neill 2017: 1618). Nowadays, we find ourselves surrounded by haptic media like the smartphone, the Apple watch and other wearables capable to "train and discipline touch in order to produce touch as a coherent communicative medium" (O'Neill 2017: 1616). Touch has undergone a transformation "into a sense capable of being stored, transmitted, and reconstructed by digital interfaces" (Parisi et al. 2017). E-textiles like the Lilypad Arduino (Buechley et al. 2008), paper-based electronics kits like Chibitronics[1] (Qi et al. 2015, 2018) and conductive touchboards[2] (e.g. De La Cruz and Bhatia 2018) rely on this, fostering an understanding of electronic touch-based cause-and-effect dualities and of more sophisticated, programmed effects.

Another line of discourse acknowledges the importance of sound for sensemaking (Bull et al. 2015) bringing forward the limitations of words for sensemaking (Wills et al. 2016), as well as their capability of developing and expressing a relationship with objects, things (Tilley 1999) and places (Thibaud 2003) through talk (Shankar 2006), and the lack thereof (Butler 2007). Polotti and Rocchesso (2008) brought this together in their focus on the creation of implicit and explicit musical knowledge backed in computer-mediated learning tools. Initiatives like Sonic Pi and Overtone are built on these insights, fostering computational learning embedded in creative, hands-on musical experiences (Aaron et al. 2016).

---

[1] https://chibitronics.com/

[2] https://www.bareconductive.com/collections/touch-board

At the desktop, computing has always been visual. Recently evolving technologies for augmented and virtual reality, however, have greatly expanded the visual sphere. They have been applied to learning in a broad range of subjects (e.g. Lu and Liu 2015), and seek to foster creativity (e.g. Yilmaz and Goktas 2017) and collaboration (e.g. Sanabria and Arámburo-Lizárraga 2017).

Also, enactment and embodiment were recognized as modes of learning (Atherton and Blikstein 2017). Antle defines embodiment as a "means how the nature of a living entity's cognition is shaped by the form of its physical manifestation in the world" (Antle 2009: 1). This view has a basis as early as Piaget (Piaget and Cook 1952), who argued that physical and mental efforts are needed to promote cognitive thinking structures in children. This has been applied to science learning (e.g. Durán-López et al. 2017). Keifert et al. (2017) pursued a similar approach through movement-based games: scientific concepts became comprehensible to children as embodied sociodramatic play, where the children imitated the behavior of, e.g., water particles. Fernaues and Tholander (2006) discuss how material artifacts in connection with physical activity like role-play can introduce children to programming concepts ("programming as performed action"). Not only do children come to a better understanding with such playful activity; programming itself becomes social, differing from the conventional screen-mouse situation, in which usually only one person makes the entry.

Making, as a concept and a practice, has the potential to combine much of the above in meaningful ways. As the act of creating tangible artifacts, making involves multiple human senses at once. It has been deployed as a means to foster learning of crafting and technology skills in combination, interlinking the digital and the physical (e.g. Rosner 2010; Peppler et al. 2016). Despite its playful character, making is described to encompass a broad set of skills, such as "cultural and material engagement, decisions around tool use, the leveraging of industrial infrastructures around materials and standards, and the crucial role of knowledge sharing and building new literacies" (Tanenbaum et al. 2013: 2604). It is concerned with individual creativity, speaking to nonlinear, hands-on learning styles (Weibert et al. 2014), collaboration (Rosner et al. 2014), and problem-solving (Lewis 2009). As *Computational Making* it combines handcrafts and the digital. It thus fosters aesthetics, creativity, construction, the visualization of multiple representations and an understanding of materials as key skills (Rode et al. 2015). It is laying a broad ground for the learning of computational skills (e.g. Juškevičienė 2020) and literacy. Urban data literacy as we explore it in our study is concerned with the ability to make sense of phenomena in the urban sphere by using, analyzing, and interpreting data and information (Schüller et al. 2019) from everyday life in the city. With our focus on crafting and making we are interested in the creative entry points to such an understanding. Our study contributes to the above laid outline of research recognizing the supportive potential of making and crafting for learning. We are exploring *crafting* and *making* as alternate methods to foster urban data literacy in young city inhabitants, as well as in those whose access to the digital urban sphere is challenged. We discuss how this can be a creative means to bring unseen city life dimensions to the fore, and to broaden access to the discourse on data and its implications for everyday city life.

Methods

This study combines principles of participatory design (Ehn et al. 2014) and action research (Kemmis et al. 2014) in a practice-based method (Wulf et al. 2018). All its workshops are conducted in the same mid-sized city in the Ruhr Area in Germany in the broader context of an initiative working to enable joint computational and media learning (Weibert et al. 2017). Situated in a neighborhood setting within that city, which is shaped by migration, our study brings together people with migration backgrounds from Afghanistan, Morocco, Poland, Romania, Russia, Syria, and Turkey (for an overview on participant numbers and recruitment, see Table 3.1). The study took place in 2019, from March to November.

Workshops

Key artifact of the workshop concept is a *city of sound*. It was collaboratively crafted by the participants of the workshops. The researchers who are also the authors of this work guided these workshops as tutors and provided help where needed. The city artifact was made from wood and equipped with programmable micro-boards. Designed in such a way, it was then used to assemble collected audio files about aspects of city life. Upon touch, these sounds could be reproduced in a certain manner from predefined spots, marked with conductive paint (see Fig. 3.1, left). The possibility to remix the sounds created a basis for discussion about abstract concepts of city life, e.g., religion and nature. Three instances of the *city of sound* were put into practice in 2019 (see Table 3.1 for an overview):

W1: This workshop was conducted as a weekend-event where children and adults from the neighborhood in focus were invited to explore the sound of the city. This was done in three steps. Children and adult participants first engaged in brainstorming and discussion about places of relevance in the neighborhood and their respective sounds. The group then set out to collect audio recordings from these places, as well as produce a city artifact from wood and paint that contained important landmarks and further details considered of relevance by the group. The collected sounds were then mapped onto the city silhouette via the touch boards and conductive paint. By touching specific places on the artifact, sounds were triggered.

Table 3.1 The workshops at a glance

	Workshop 1	Workshop 2	Workshop 3
Topic	Environment/nature	Religion/faith	Religion/faith
Location	Community center	Municipal event center	Christian church center
Participants	20/open recruitment process	~20/open participation over the course of 1 month	~20/open participation over the course of 2 months

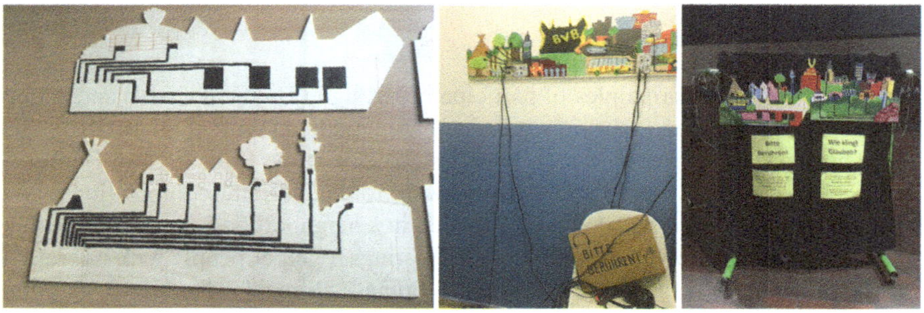

Fig. 3.1 The city of sound was crafted from wood. Equipped with programmable micro-boards, it could produce sound upon touch of designated spots which were marked with conductive paint

W2: A pre-made wooden city silhouette became a part of an exhibition in a municipal event center downtown. Publicly accessible to its audience for 1 month it invited the exploration of the "sound of faith." People of all ages could contribute sounds that were of relevance to them in two ways, (1) by joining exhibition events in person, and (2) by sending audio files via e-mail to be added to the city.

W3: This workshop invited the exploration of the "sound of faith" in a Christian church center. The crafted city artifact was publicly accessible, and youth as well as adults engaged with it at times of services and church events for 2 months.

Data Collection and Analysis

The study data presented here consists of the field notes, images and the artifacts resulting from the workshops. We documented all workshop activity including verbatim feedback of participants and attendees in short session notes, which we extended to full field notes afterward. The authors conducted the workshops in cooperation with residents from the field and provided guidance as tutors, as required. For the analysis (Fereday and Muir-Cochrane 2006) we were interested in technical and computational understanding that participants showed, and how these were used to make sense of abstract concepts of urban everyday life.

Findings and Their Discussion

Our qualitative and thematic analysis yielded *materiality, place,* and *diversity of learners* as the main themes. In the following, we discuss our findings for each concept in turn.

Materiality

Crafting activity was the initial means to structure and approach the respective topic. This was the case in W1 when the young and adult participants set out to detail the roughly pre-shaped wooden city artifact. By detailing the wooden elements, discussion was fostered exploring what's a landmark, and what places and things of (individual) importance should further be added to the city. A large, high-rise residential building complex was proposed by one of the women as one of those details: *"Everything from my everyday life is right there: family, the school of my kids, shopping..."* She discussed with her friend how to paint the building and ended up cutting an image of the building from a local newspaper, gluing it to the wood, *"to make it look realistic."*

Through these material, wooden and paper details, "the sound of the city" was then explored, e.g. when a teenager said he wanted to add a "cool car." His focus widened from the initial material decisions to make (what color, what shape) toward a broader view that included research on the sound of the engine, and discussion with two other boys, where to place the car in the city to have it *"being seen."* Another example was a boy who initially associated the sound of fighting on the street with the high-rise residential building complex he lived in. The proposal to add this sound initially caused irritation in the group, which quickly resolved to a serious conversation, where three of the women pointed to this detail, saying how the boy was right, and that there were frequent fights and violence and police operations in this area. *"If we add it to the city, we point to this problem, and people become aware."* Two women discussed where to place birds: *"In the playground maybe? They do not really have a place, have they?"*

In the case of W3, crafting as an activity fostered discussion across generations. An example for this is the conversation among a senior and a teenager, who were talking about how modes of expression had technically evolved while exploring the functionalities of the touchboard and how its size enabled an unobtrusive integration into the artfully crafted and painted city artifact. The senior had knowledge in woodcarving and audio broadcasting technology; he voiced his fascination with the individualized sound experiences that the city artifact provided to him in a programmed manner. The teenager had some coding skills; he was intrigued to apply these skills to his fascination with religious diversity (How does one translate faith and urban religious experience into code, combined with crafted wood and some paint?). Such engagements with touch-based cause-and-effect dualities as well as sophisticated, programmed effects deploying touch as "coherent communicative medium" (O'Neill 2017: 1616) prolong earlier findings on the supportive nature of touch and tangibles for learning (e.g. De La Cruz and Bhatia 2018) to the realm of data literacy in an urban context.

Diversity of Learners

By enabling sound as a part of the city design, reflection was fostered on the different shapes an information can have—a basic element of data literacy. By audio recording and editing sounds of their neighborhood, children and adults in W1 came to think and discuss their meanings: Is this noise? Music? Nature? Do I want this? Teams in the workshop collected playground sounds, a rap song a pedestrian had spontaneously performed for them, construction site noise, and cars rushing by, thus assembling the ingredients for what Mattern has described as "site-based experience" and "sensory engagement" (Mattern 2021:70). A recording of birds chirping was cherished by one woman, noting this was a sound that was easily missed in the overall soundscape of the city.

All three instances of the city supported such development of data literacy among a broad spectrum of child and adult learners by having the sounds and their combination programmed. In W1 the city was crafted in a way that a single headset was connected to several touchboards (see Fig. 3.1, middle), enabling what Hansen has coined to be "seeing through the hand" (Hansen 2006: 71): the creation of a sound collage on the spot by touching multiple places on the city in combination. Exploring "the sound of faith" as part of an exhibition in the public center downtown in W2, the touchboards were programmed in such a way that sounds were randomly assigned to the different places in the city and played upon touch (see Fig. 3.1, right). This created a soundscape that had the broad variety of religious sounds being played all over the city. A Muslim woman especially valued this in the exhibition. She was noting how this type of coding enabled the call of the imam to be heard at the city hall right in the heart of the city, whereas in the reality that she experienced, it was rather pushed to backyards and industrial areas in the city outskirts. A group of four male attendees noted in W3 how with this type of script, the image of their city as a home for soccer was supported, with stadium chants being heard everywhere in town (*"Soccer really is some kind of a religion here, too."*). Both incidents can be read as examples of an evolving literacy that is not only capable of differentiating different types of urban data, but also able to recognize how these can be turned into a statement with the crafting to include and express "site-based experience" (Mattern 2021: 70).

Place

In all of this, we saw notions of place to be a powerful factor—both with regard to topic, as well as to how the crafting and making unfolded. The city of sound in W1 was crafted in a community center, a surrounding the people participating were comfortable with. The place provided a familiar basis to engage with the topic, and children and adults did not hesitate to contribute personal views and perspectives. As a municipal event center, the place of W2 provided a more public surrounding.

This was supportive to engagement with the artifact and fostering discussion, but in a more formal manner. The difference could most vividly be observed in the reaction of some of the participants of W1: two women and their children were enthusiastic about the activity and spontaneously announced that they would attend future workshops with this topic as well—however, in the municipal event center they did not feel comfortable in the same way, so they stepped back from their initial plan. As a Christian church center, the place of W3 came with a predefined main audience, and this was reflected in how the city artifact was received. Discourse revolved around faith and its meanings in everyday life in the neighborhood—but with an emphasis on Christianity.

Conclusion

We have explored how making and crafting can be a means to foster the building of urban data literacy—a skillset that is concerned with the ability to make sense of phenomena in the urban sphere by using and interpreting data and information. We have discussed in our works with a crafted *city of sound* installation, how this can bring unseen city life dimensions to the fore, and how it is linked to digital literacy and creative data literacy: To the participants of our workshops the creativity that the crafting and making provided turned out to be a means to figure out the shape and value of an information (as seen, e.g., in the collection and processing of urban sounds in W1). The findings from our study indicate that a tactile and artistic approach to such data and its meaning in the city provides creative entrances to a topic that by speaking also to the human senses such as touch and sense of hearing go beyond established ways of sensemaking through speaking, writing, and calculating. The joint crafting around the specific data and information as enabled in our workshop activity fosters their discussion across communities, thus extending the collaboration and interaction with others that Kafai and Burke (2014) talked about earlier to a data literacy and digital literacy context. This was the case with participants in W2, who recognized the possibility to program experiences of religious exclusion into the city artifact, or in W3, where the crafting enabled collaboration and reflection across generations. Parisi, Paterson, and Archer have earlier recognized a transformation of touch "into a sense capable of being stored, transmitted, and reconstructed by digital interfaces" (Parisi et al. 2017)—in relation to the crafting of the *city of sound*, we could see it being explored as a means that can convey programmed messages to a previously identified audience thus actively fostering a sense for digital literacy and participation in an urban context. Not only did the participating children, youth, and adults recognize the joy that can be inherent to creating a beautiful or cool artifact—they discussed notions of an audience for such creation, as well as possibilities to "make a statement" with this. Such activity can thus be a creative means to add to the data basis used to legitimize urban design choices and foster a broader degree of participation.

Acknowledgments Funded by the German Research Foundation (DFG) in the context of the Collaborative Research Centre Media of Cooperation (262513311 – SFB 1187 and GRK 1769). Gefördert durch die Deutsche Forschungsgemeinschaft (DFG) – Projektnummer 262513311 – SFB 1187 und GRK 1769.

References

Aaron S, Blackwell AF, Burnard P (2016) The development of Sonic Pi and its use in educational partnerships: co-creating pedagogies for learning computer programming. J Music Technol Educ 9(1):75–94

Albino V, Berardi U, Dangelico RM (2015) Smart cities: definitions, dimensions, performance, and initiatives. J Urban Technol 22(1):3–21

Antle AN (2009) LIFELONG INTERACTIONS - embodied child computer interaction: why embodiment matters. Interactions 16:27

Antle AN, Wise AF, Nielsen K (2011) Towards utopia: designing tangibles for learning. In: Proceedings of the 10th international conference on interaction design and children. ACM

Atherton J, Blikstein P (2017) Sonification blocks: a block-based programming environment for embodied data Sonification. In: Proc. of the conference on interaction design and children. ACM

Barr V, Stephenson C (2011) Bringing computational thinking to K-12: what is involved and what is the role of the computer science education community? ACM Inroads 2:48–54

Boy GA (2013) From STEM to STEAM: toward a human-centred education, creativity & learning thinking. In: ECCE '13: proceedings of the 31st European conference on cognitive ergonomics, vol Article No. 3, pp 1–7. https://doi.org/10.1145/2501907.2501934

Bröring A, Remke A, Lasnia D (2011) SenseBox–a generic sensor platform for the web of things. In: International conference on Mobile and ubiquitous systems: computing, networking, and services. Springer, pp 186–196

Buechley L, Eisenberg M, Catchen J, Crockett A (2008) The LilyPad Arduino: using computational textiles to investigate engagement, aesthetics, and diversity in computer science education. In: Proceedings of the SIGCHI conference on human factors in computing systems, pp 423–432

Bull M, Back L, Howes D (eds) (2015) The auditory culture reader. Bloomsbury Publishing

Butler T (2007) Memoryscape: how audio walks can deepen our sense of place by integrating art, oral history and cultural geography. Geogr Compass 1(3):360–372

Chawla L (2002) "Insight, creativity and thoughts on the environment": integrating children and youth into human settlement development. Environ Urban 14(2):11–22

Dewey J (1923) Democracy and education: an introduction to the philosophy of education. Macmillan, New York

De La Cruz S, Bhatia A (2018) Paper piano: making circuits with everyday things. In: Proceedings of the 17th ACM conference on interaction design and children, pp 521–524

Durán-López E, Rosenbaum LF, Iyer GV (2017) Geometris: designing collaborative mathematical interactions for children. In: Proc. of the Conference on Interaction Design and Children. ACM

DiSessa AA (2001) Changing minds: computers, learning, and literacy. MIT Press

Dourish P (2004) Where the action is: the foundations of embodied interaction. MIT Press

Ehn P, Nilsson EM, Topgaard R (2014) Making futures: marginal notes on innovation, design, and democracy. The MIT Press, p 392

Fauville G, Lantz-Andersson A, Säljö R (2014) ICT tools in environmental education: reviewing two newcomers to schools. Environ Educ Res 20:2. https://doi.org/10.1080/13504622.2013.775220

Fereday J, Muir-Cochrane E (2006) Demonstrating rigor using thematic analysis: a hybrid approach of inductive and deductive coding and theme development. Int J Qual Methods 5(1):80–92

Fernaeus Y, Tholander J (2006) Designing for programming as joint performances among groups of children. Interact Comput 18:1012–1031
Gabrys J (2016) Program earth: environmental sensing technology and the making of a computational planet. University of Minnesota Press
Guzdial M (2008) Education paving the way for computational thinking. Communications 51:25–27
Hamidi F, Baljko M (2017) Engaging children using a digital living media system. In: Proceedings of the 2017 conference on designing interactive systems, pp 711–723
Hansen M (2006) Bodies in code: interfaces with digital media. Routledge, New York
Kathrin Hörschelmann & Lorraine van Blerk (2013) Children, youth and the city. Routledge, London
Holloway SL, Valentine G (2000) Children's geographies. Playing, living, learning. Routledge, London
Holmquist LE, Redström J, Ljungstrand P (1999) Token-based access to digital information. In: International symposium on handheld and ubiquitous computing. Springer, Berlin, pp 234–245
Juškevičienė A (2020) Developing algorithmic thinking through computational making. In: Data science: new issues, challenges and applications. Springer, pp 183–197
Kafai YB, Burke Q (2014) Connected code: why children need to learn programming. MIT Press
Keifert D, Lee C, Dahn M, Illum R, DeLiema D, Enyedy N, Danish J (2017) Agency, embodiment, & affect during play in a mixed-reality learning environment. In: Proc. of the 2017 Conf. On interaction design and children, pp 268–277
Kemmis S, McTaggart R, Nixon R (2014) The action research planner: doing critical participatory action research. Springer
Kunze J (2020) Data literacy in the smart city. Geoforum Perspektiv 19(35). https://doi.org/10.5278/ojs.perspektiv.v19i35.6423
Lewis T (2009) Creativity in technology education: providing children with glimpses of their inventive potential. Int J Technol Des Educ 19(3):255–268
Lockwood J, Mooney A (2017) Computational thinking in education: where does it fit? A systematic literary review. arXiv preprint arXiv:1703.07659
Lohr S (2009) For Today's graduate, just one word: statistics. The New York Times
Lu S-J, Liu Y-C (2015) Integrating augmented reality technology to enhance children's learning in marine education. Environ Educ Res 21(4):525–541
Marshall P, Cheng P, Luckin R (2010) Tangibles in the balance: a discovery learning task with physical or graphical materials. In: Proc. of the Int. Conf. On tangible, embedded & embodied interaction
Mattern S (2021) A City is not a computer: other urban intelligences. Princeton University Press
McGrath B (2001) "A problem of resources": defining rural youth encounters in education, work & housing. J Rural Stud 17(4):481–495
McLuhan M (1994) Understanding media: the extensions of man. MIT Press
Merleau-Ponty M (2013) Phenomenology of perception. Routledge, New York
O'Malley C, Fraser DS (2004) Literature review in learning with tangible technologies. ffhal-00190328f
O'Neill C (2017) Haptic media and the cultural techniques of touch: the Sphygmograph, Photoplethysmography and the apple watch. New Media Soc 19(10):1615–1631
Panelli R, Punch S, Robson E (eds) (2007) Global perspectives on rural childhood and youth: young rural lives. Routledge
Papert S (1980) Mindstorms: children, computers, and powerful ideas. Basic Books
Parisi D, Paterson M, Archer JE (2017) Haptic media studies. New Media Soc 19/10:1513–1522
Peppler K, Halverson ER, Kafai YB (2016) Makeology: makers as learners. Routledge
Piaget J, Cook MT (1952) The origins of intelligence in children. Int. Universities Press, New York
Pink S (2015) Doing sensory ethnography. Sage, Los Angeles
Polotti P, Rocchesso D (2008) Sound to sense, sense to sound: a state of the art in sound and music computing. Logos, Berlin

Qi J, Buechley L, Huang AB, Ng P, Cross S, Paradiso JA (2018) Chibitronics in the wild: engaging new communities in creating technology with paper electronics. In: Proceedings of the 2018 CHI conference on human factors in computing systems. ACM, pp 1–11

Qi J, Huang AB, Joseph Paradiso J (2015) Crafting technology with circuit stickers. In: Proceedings of the 14th international conference on interaction design and children. ACM, pp 438–441

Rode JA, Weibert A, Marshall A, Aal K, von Rekowski T, El Mimouni H, Booker J (2015) From computational thinking to computational making. In: Proc. of the 2015 ACM Int. joint conference on pervasive and ubiquitous computing. ACM, pp 239–250

Rodgers S, Ploderer B, Brereton M (2019) HCI in the garden: current trends and future directions. In: Proc. of the 31st Australian conference on human-computer-interaction, pp 381–386

Rosner DK (2010) Mediated crafts: digital practices around creative handwork. In: CHI'10 extended abstracts on human factors in computing systems, pp 2955–2958

Rosner DK, Lindtner S, Erickson I, Forlano L, Jackson SJ, Kolko B (2014) Making cultures: building things & building communities. In: Proc. of the companion publication of the 17th ACM conference on computer supported cooperative work & social computing, pp 113–116

Sanabria JC, Arámburo-Lizárraga J (2017) Enhancing 21st century skills with AR: using the gradual immersion method to develop collaborative creativity. Eurasia J Math Sci Technol Educ 13(2):487–501

Schüller K, Busch P, Hindinger C (2019) Future Skills: Ein Framework für Data Literacy – Kompetenzrahmen und Forschungsbericht. Arbeitspapier Nr. 47. Hochschulforum Digitalisierung, Berlin

Sefton-Green J, Nixon H, Erstad O (2009) Reviewing approaches and perspectives on "digital literacy". Pedagog Int J 4:107–125

Shankar S (2006) Metaconsumptive practices and the circulation of objectifications. J Mater Cult 11(3):293–317

Soro A, Brereton M, Dema T, Oliver JL, Chai MZ, Ambe AMH (2018) The ambient birdhouse: an IoT device to discover birds and engage with nature. In: Proc. of the 2018 CHI conference on human factors in computing systems. ACM, pp 1–13

Tanenbaum JG, Williams AM, Desjardins A, Tanenbaum K (2013) Democratizing technology: pleasure, utility and expressiveness in DIY and maker practice. In: Proc. of the 2013 CHI conference on human factors in computing systems. ACM, pp 2603–2612

Thibaud J-P (2003) The sonic composition of the city. In: Bull M, Back L (eds) The auditory culture reader. Berg, Oxford, pp 329–342

Tilley C (1999) Metaphor and material culture. Blackwell Publishing

Ufan L (1973) On the hand. In Glenn Adamson (2010). The craft reader. Berg, New York, pp 548–551

Ullmer B, Ishii H (2000) Emerging frameworks for tangible user interfaces. IBM Syst J 39(3.4):915–931

Ullmer B, Ishii H, Jacob RJ (2005) Token+ constraint systems for tangible interaction with digital information. ACM Trans. Comput Hum Interact (TOCHI) 12(1):81–118

Weibert A, Marshall A, Aal K, Schubert K, Rode J (2014) Sewing Interest in E-Textiles: Analyzing Making from a Gendered Perspective. In: Proceedings of the 2014 Conference on designing interactive systems. ACM, pp 15–24

Weibert A, Randall D, Wulf V (2017) Extending value sensitive design to off-the-shelf technology: lessons learned from a local intercultural computer Club. Interact Comput 29(5)

Wills WJ, Dickinson AM, Meah A, Short F (2016) Reflections on the use of visual methods in a qualitative study of domestic kitchen practices. Sociology 50(3):470–485

Woodward S (2019) Material methods: researching and thinking with things. Sage, Los Angeles

Wolff A, Gooch D, Cavero JJ, Montaner UR, Kortuem G (2016) Creating an understanding of data literacy for a data-driven society. J Community Inform 12(3)

Wulf V, Pipek V, Randall D, Rohde M, Schmidt K, Stevens G (eds) (2018) Socio-informatics. Oxford University Press

Xie L, Antle AN, Motamedi N (2008) Are tangibles more fun? Comparing children's enjoyment and engagement using physical, graphical and tangible user interfaces. In: Proceedings of the 2nd International Conference on Tangible and Embedded Interaction, pp 191–198
Yilmaz RM, Goktas Y (2017) Using augmented reality technology in storytelling activities: examining elementary students' narrative skill and creativity. Virtual Reality 21(2):75–89

Chapter 4
Scanning for Scams: Local, Supra-national, and Global Events as Salient Contexts for Online Fraud

Kristjan Kikerpill

Introduction

Regardless of the verbs we use when describing our online presence, e.g., online banking, dating, gaming, or shopping, these phenomena boil down to a series of requests and responses exchanged between connected devices and servers across the globe (Kikerpill 2021a). Digital spaces and places so-called are but informational nodes signalling an active human presence and waiting to be acted upon (Maggi 2014). Communication in such environments (see Snowdon et al. 2001) connects us to other people, and the digital traces of their activities, through texts and graphical interfaces. All of our actions and interactions in mediated environments are, in fact, communicative acts and accompanying interpretations carried out via open channels and with the help of various media. In other words, people's mediated presence always constitutes action-as-communication (Kikerpill 2021a).

Since crime is a socially constructed phenomenon (Posick 2018), it goes wherever people go. Thus, the opportunities and (near-)immediate access provided by modern information and communications technologies are not always used towards positive or legal ends, but are employed instead to perpetrate cybercrimes. Put differently, some action-as-communication in mediated environments constitutes crime-as-communication (Kikerpill 2021a). In recent years, both the financial and psychological harms caused by cybercrime victimisation are on the rise (PurpleSec 2021), and especially from online scams and frauds (Button and Cross 2017). The fact that approximately 99% of cybercrime threats require human interaction, e.g., opening attachments in emails or following links (Proofpoint 2019), to be

K. Kikerpill (✉)
Institute of Social Studies, University of Tartu, Tartu, Estonia
e-mail: kristjan.kikerpill@ut.ee

D. Radovanović (ed.), *Digital Literacy and Inclusion*,
https://doi.org/10.1007/978-3-031-30808-6_4

successful makes the recipient of a mediated crime attempt the person best positioned to mitigate relevant risks and ensure their own safety from victimisation (Kikerpill 2021b).

Hence, the first step towards preventing harm from the modern social menace of cybercrime relies on a simple, but important understanding: in mediated environments, there is plenty of communication without crime, but no crime without communication. As perpetrators often manipulate the communication underlying their cybercriminal activities, i.e., socially engineer the messages in their online attacks (Hadnagy 2018; Hatfield 2018), learning to distinguish between criminal and non-criminal communication is essential. The most common general form of social engineering attacks in mediated environments is broadly referred to as "phishing" (Khonji et al. 2013), although numerous variations of phishing exist, e.g., vishing (voice phishing) and smishing (text message phishing) (Hong 2012). An important part of the larger effort of countering cybercrime comes from understanding how criminal actors exploit salient social contexts within the content of the scams and frauds they disseminate. Given that contexts have both an interpretive and a constitutive dimension (Rigotti and Rocci 2006), they help people interpret incoming messages, but can also be used to create messages that fit specific social expectations (Carter 2015). In cybercriminal endeavours, this means exploiting salient social circumstances to craft crime messages, which are more meaningful for the recipient due to a shared "lived experience" (Kikerpill 2021a).

To explore and illuminate how salient social contexts can spur online scams, and how awareness about these connections can contribute to preventing victimisation from fraud, we apply the *mazephishing* framework (Kikerpill and Siibak 2021a) to study specific events on the local, supra-national and global level. The *mazephishing* framework comprises three primary components: the social context from which specific scam messages obtain their salience, e.g., the COVID-19 pandemic (Kikerpill and Siibak 2021b) or natural disasters such as forest fires (Taodang and Gundur 2022), the media or channels used to circulate the scam messages (see above: Hong 2012), and the influencing techniques employed in the actual scam messages (Lawson et al. 2020; Kikerpill 2021a; Steinmetz et al. 2021). In this chapter, the focus will primarily be on the social context element, because we are only now beginning to learn about the true importance of social context in cybercrimes that require human interaction (Verma et al. 2018; Norris et al. 2019; Montañez et al. 2020; Kikerpill and Siibak 2021a; Steinmetz et al. 2021; Taodang and Gundur 2022), and how this emerging knowledge can be used in digital literacy education for the purposes of fraud avoidance. Scholarship and education in this area is paramount if we want to move towards dismissing the entrenched discourse of the "deficient user" that currently dominates cybersecurity discussions (Klimburg-Witjes and Wentland 2021).

Background and Approach

Since fraud is a crime of interaction (Harrington 2012: 396), both its offline and online manifestations are always rooted in and dependent upon communication (Kikerpill 2021a). While all cybercrime depends on communications technology, the concurrent communicative aspects of the same crimes are often overlooked if not diminished in lieu of more technical discussions (Kikerpill 2021a), e.g., "to a computer scientist, the solution to a bug is often just more computer science" (Borel 2018). Taking this into account, the chapter decidedly focusses on the equally important communicative and interpretive underpinnings of cybercrime by presenting a series of examples (Simons 2014) of specific salient events or circumstances that have enabled criminals to use the entailing social context as input for their socially engineered fraud messaging. Where contexts are not primarily created within a fraudulent interaction, e.g., in longer running online dating and romance scams (Carter 2021) or cold-call type "one-off" fraud attempts such as phishing attacks (Khonji et al. 2013; Atkins and Huang 2013; Kikerpill and Siibak 2019), criminals can decrease their deviant workload and increase the credibility of the crime messages, by relying on events or circumstances that are important in a geographically, culturally or temporally restricted, semi-open or open manner.

Acknowledging that not all events are of equal importance for different communities in various parts of the world at any given moment, we use cases with local (geographically and temporally restricted), supra-national (geographically and culturally semi-open, temporally restricted) and global (culturally and geographically open, temporally semi-open) significance. The categories of geographical, cultural and temporal openness and/or restrictions are used as guidelines for better understanding the connection between social occurrences and fraud proliferation, including why some types of scam content may be relevant for some and not others. The reasoning is that interpretations of scam believability can depend on where we live, which cultural practices we observe and what we consider as desirable or necessary at any given moment (Kikerpill 2021a). The *mazephishing* framework was chosen as a lens for exploring the aforementioned categories because it provides a structured backdrop with respect to what people should look for in scams in general, i.e., the (social) timing of particular scams, the relevance and comprehensibility of scam messages depending on the current "lived experience" of a person as well as how we engage with modern mediated environments in general.

While the chapter does not directly include temporally unrestricted cases, these would mainly involve malicious exploitations of the human experience and people's vulnerabilities rather than the amplification provided by any single event or specific circumstances (Kikerpill 2021a), e.g., as it often occurs in scams perpetrated in the context of intimate relationships and romance (Carter 2021). Yet, it must be noted that while these opportunities are available to scammers without particular temporal restrictions, the prevalence of romance scams is also known to increase during Valentine's Day (Fowler 2022). Even so, the examples in this chapter focus on events or circumstances with an element of temporal restriction to also explore the idea of "criminal event calendars", i.e., how (cyber)criminals may be perceiving, or telling, time in accordance with specific opportunities for criminal exploitation based on salient social contexts.

On the local level, we present a case study of scams circulated during the respective tax seasons in Estonia and the United States. For the supra-national level, we provide examples from widely recognised commercial sales events, i.e., Amazon Prime Day and Black Friday. For the global level, we chose the current phenomenon of gaming console unavailability and restocking issues that have been caused by a shortage in microchips required for the production of said consoles. These examples represent (1) instances where an obligation necessitates certain practices, and the context of this obligation creates opportunities for scammers; (2) instances where cultural and commercial developments have created certain opportunities for scammers, and (3) instances where a combination of unexpected circumstances create opportunities for scammers.

It is important to note that the examples presented in the chapter are not geared towards bringing about or recommending substantive changes in the events or circumstances as such – which, as will become clear, would be very difficult if not entirely impossible – but are meant as illustrative examples on how the realities of the social world become reflected in mediated crimes, and how being aware of these connections can aid in avoiding becoming a victim of fraud.

Online Scam Ecosystem During Tax Season in Estonia and the United States

In this chapter, the previously mentioned categories of geographically and temporally restricted contexts mean that a similar or identical event occurs on different set dates or date ranges in different countries, which makes it possible to explore how the event or circumstances impact the dissemination of contextually fitting scams. For instance, the so-called tax season begins in January in the United States, but in mid-February in Estonia, with respect to private individuals' tax declarations. Given the vast differences in tax filing complexity between the two countries (e-Estonia 2021), the following exemplifies how opportunities for fraudulent offers made by scammers may differ in scope and intensity.

The starting point for tax-related frauds comes from the importance of the institution as such and people's willingness to pay their taxes. Attitudes towards paying taxes vary significantly across different countries, where tax morale is influenced by numerous factors such as cultural differences and trust in one's government (Torgler and Schneider 2007). In Estonia, 91% of people consider paying taxes their essential obligation (ETCB 2021), 98% of personal income declarations are made electronically (e-Estonia 2022) and the Estonian Tax and Customs Board's e-tax system is viewed as the most convenient public service being offered (Kantar Emor 2020). Furthermore, the average personal income tax declaration takes approximately three to five minutes to file (Work in Estonia 2022), which makes tax compliance easy. In contrast, the tax preparation and filing process in the United States can take approximately 13 hours for an individual (Kessler 2013), and about 44% of Americans are bothered "a lot" by the complexity of the tax system (Pew Research 2015). From the perspective of scammers, who are known to be opportunistic in their exploitation of people's vulnerabilities (Kikerpill and Siibak 2021b), the more complexity a particular system presents, the more opportunities there are for interjecting bogus offers for seemingly relevant services, including for the speeding up or simplification of the process.

Following from the above, there were only a very limited number of tax season scams available for further analysis with respect to Estonia. With the exception of 2018, there was at least one reported tax scam from 2014 to 2020 and the time of reporting ranged from late January to mid-March. The outlier was a tax refund scam reported at the end of December (Sobak 2014), which requested people to submit their credit card information for an expedited tax return. As also noted in the relevant scam report (Sobak 2014), the circulated fraud message was mistimed by the criminals, because personal income tax declarations are filed starting from February 15. Hence, examples from other years appear on and around February 15 and in March. There were two main types of scams disseminated: phishing emails that request the recipient to provide additional information to receive their tax refund quicker (Pihlak 2017) or which provide a link that leads the recipient to a faked website of the local tax authority for the purposes of entering one's credit card number and the relevant security code (Raamatupidaja 2016). Interestingly, a scam circulated in 2016 (Rapp 2016), which used bad Estonian and notified recipients that the tax authority was unable to process their respective tax refund and, thus, requires additional information, also promised the return to be made in Estonian kroons, i.e., the currency used in Estonia prior to 2011 and the Euro. Hence, not only can temporally restricted scams be noticed and reported due to mistimed dissemination and poor use of local language, but also when the scams fail to take into account local changes and social context. From a technical perspective, since credit card numbers and security codes are only used to initiate payments (Walter 2019), and not to receive them, providing the tax authority with one's relevant respective information lacks purpose entirely.

In comparison with the Estonian examples, the scam ecosystem of the US tax season is a completely different phenomenon. Firstly, tax season scams are so widespread in the United States each year that it has become commonplace to release general warnings beforehand (Rafter 2022). In contrast, the scam reports were few in Estonia and reported only after the scams actually occurred. Furthermore, the complexity of the tax filing process (e-Estonia 2021) reveals that tax preparation services are common in the United States, but virtually unheard of for private individuals in Estonia. As mentioned previously, the complexity of a process, i.e., the number of steps a person has to take in order to complete the process, presents opportunities for scammers to interject bogus offers or threats. Thus, it is not surprising that one of the more common types of tax season scams in the United States relates to fraudulent tax preparation services (Rafter 2022). A related issue concerns taxpayer advocate scams in which recipients receive a call and are asked for personal information that would allow the perpetrators to successfully commit identity theft (Rafter 2022). Provided that tax advocates aid taxpayers with the more difficult tax issues, this further shows how the complexity of a process can increase the variety of scams it potentially enables. In comparison, since the majority of Estonian personal income declarations are pre-filled and the process takes only some minutes in the official online environment of the local tax authority (see e-Estonia 2021), a significant number of scam opportunities are avoided through this solution.

Digital Hallmark Holidays Mark a Rise in Scams: Amazon Prime Day and Black Friday

Originating from the United States, the term "hallmark holidays" broadly refers to the celebration or observance of dates primarily for commercial purposes. In the digital sphere, this has come to include "commercial holidays" such as Amazon Prime Day, which has been in effect since 2015 to celebrate the 20th anniversary of the company Amazon (Johnston 2022), as well as Black Friday that arrives yearly at the end of November. Commercial events like Amazon Prime Day and Black Friday are geographically and culturally semi-open due to the increasing reach of Amazon's activity, and the adoption of Black Friday sales events in countries other than the United States Since crime, including scams, goes where people go (Posick 2018; Kikerpill 2021a), geographically and culturally semi-open contexts for scams can expand over time insofar as new communities take up the practice of "following" certain dates or events.

For instance, Black Friday was historically restricted primarily to the United States (Marcos 2021), but the opportunity for financial gain from set-date steep sales has made the observance of this commercial event spread to other countries via globalisation (Dumoulin 2019). Although the aforementioned Black Friday sales events take place in different countries, each location still observes the original date for the event, i.e., late November, which makes it temporally restricted. Similar to Black Friday, the initial reach of Amazon Prime Day has also expanded in unison with the company's increasing sphere of activities – while Prime Day began as a 24-h sales event that included 9 countries, it has since grown into a 48-h event spanning 20 countries (Johnston 2022). The dates of such temporally restricted but culturally and geographically semi-open sales events are, therefore, prime targets for scammers to present their fraudulent offers alongside legitimate offers from stores and online merchants. It is important to note here that Amazon Prime Day has usually occurred in the month of June or July, but took place in October in 2020 due to the COVID-19 pandemic (Johnston 2022). The importance of such a shift is revealed in the corresponding warnings circulated in the media concerning "Amazon Prime scams" (Tompor 2021; Whitney 2021), i.e., the scams follow the dates of an event even if the date is changed due to exceptional circumstances, which also provides some support for the notion of so-called criminal calendars.

Furthermore, although Amazon Prime Day is still only geographically semi-open in its reach, the significance of the sales event is expansive enough to prompt "preparatory" scams (Bolster Blog 2021), i.e., scams that are perpetrated even before the actual event begins. For instance, these preparatory scams include offers for early deals as well as attempts to entice incoming users to become Amazon members, including with various fraudulent offers for coupons and discounts, and set up their respective payment accounts (Bolster Blog 2021). Given the sales frenzy of the actual event, perpetrators are able to intensify their otherwise regular efforts and exploit people with ruses that are built on non-existent problems with a person's Amazon account, on bogus payment and shipping receipts that are meant to make the person submit additional personal information, as well as on the "verification" of payment methods used in an Amazon purchase (ITRC 2019). Moreover, even after the event-proper has passed, the online sales ecosystem's general reliance on product reviews provides scammers with a further opportunity for perpetrating fraud (ITRC 2019), i.e., criminals are able to make bogus monetary offers in return for writing reviews that are only a smoke-screen for stealing a person's personal and payment information.

Scammers' approach to Black Friday sales is similar to those employed with respect to Amazon Prime Day. Potential buyers are presented with offers that are "too good to be true", asked for personal information or payment details under the guise of fraudulent delivery messages, and lead to enter their payment information into very real-looking fake websites of online merchants (Smith and Aguilar 2021; Osborne 2021). Of note with Black Friday events is their increased sphere of influence due to globalisation (Dumoulin 2019), which is also represented in how the scam ecosystem of fake websites is created alongside the efforts of legitimate vendors trying to take advantage of the sales dates (Bischoff 2020). As Bischoff (2020) showed, the registration of new websites skyrockets in the period preceding Black Friday and Cyber Monday, which is a sibling event to the former, and these websites are spread out globally. Thus, what was historically an event primarily observed in the United States, has expanded throughout most of the world because of its potential for bringing in buyers that are looking for discounts and probably also making materialistic preparations for Christmas. In effect, the salience of the Amazon Prime Day and Black Friday events comes from their regular and relatively reliable occurrence each year, which allows scammers to prepare crime messages, lures and dissemination tactics beforehand. Moreover, the further the occurrence and the legitimate exploitation of such events reaches, the more salient context "room" there is for fraudsters to operate (Dumoulin 2019; Bischoff 2020; Osborne 2021). Even though the aforementioned sales events are temporally restricted to one or two days, the events carry enough significance and have created certain expectations, for buyers that the pre- and post-event periods are also marked in the respective "criminal calendars". Moreover, since Black Friday is itself a commercial prelude to Christmas, the last months of the year are dotted with legitimate commercial events that are as busy for scammers as they are for retailers.

Different from the first example of tax season, which focussed on how certain demands for services are created by specific obligations that people are subject to, the criminal exploitation of commercial holidays often comes down to criminal actors "piggybacking" on gain-based incentives already present for those interested in discounts and deals. As long as people are sufficiently incentivised to engage in practices that involve transfers of funds, the source of the specific demands for goods and services, i.e., whether legally prescribed or culturally created, is less important in mediated fraud.

Non-existent Stocks Can Never Run Out: The COVID-19 Microchip Shortage and Gaming Console Scams

Scammers tend to create their main ruse based on one of two communicative approaches, i.e., either a gain-based "Good Samaritan", i.e., offering items or services currently in demand, or a loss-based "Shock and Awe" approach, i.e., threatening to cause financial or reputational harm to persons (Kikerpill and Siibak 2021b). Since the social context of frauds has been shown to significantly impact the content of crime messages circulated to the public (Kikerpill 2021a; Taodang and Gundur 2022), the content of such scams, in turn, also reflects the opportunities that the particular context allows for. For instance, scams disseminated in the first four months of the COVID-19 pandemic relied more on a gain-based approach (Kikerpill 2021a), because the pandemic circumstances themselves better facilitated fraudulent offers of potential gain more so than threats of loss. These included bogus offers for difficult-to-obtain personal protective equipment, various untested cures and remedies and even vaccines (Naidoo 2020; Kikerpill and Siibak 2021a). Therefore, when the social context created by an event or salient circumstances is more open to offering recipients something that they need or want rather than threatening to take away something that people already have, then this notion can be expected to be reflected in the types of online scams being circulated.

Following from the above, and considering how much of today's world "runs" on microchips, i.e., smartphones, laptops and even cars (Feder 2021), the final example provides an initial glimpse into what happens in terms of online fraud if there is suddenly a shortage in the supply of such objects of desire or need. Here, the culturally and geographically open social context for scams originally emerged from a combination of at least two important developments during the COVID-19 pandemic: the increased number of people working from home and using smart devices for work, school and entertainment (Vargo et al. 2021) as well as the issues with and limits to the process of manufacturing microchips (Kamasa 2021). Items such as gaming consoles fall under both of the aforementioned categories, i.e., consoles require microchips and are an increasingly important part of home entertainment (Muriel and Crawford 2018). Thus, when the newest Xbox and PlayStation 5 released only two days apart in November 2020, it was a global cultural event that occurred in the midst of the COVID-19 pandemic (Frank 2020). Even though Sony, i.e., the manufacturer of PlayStations, did not expect the COVID-19 pandemic to derail the new console's release plans (Powell 2020), keeping the gaming consoles in stock, including in online stores, was highly problematic from the beginning (Smith 2020). As already shown previously, deficits concerning in-demand products or services are a quintessential opportunity for scammers to defraud people (Kikerpill and Siibak 2021b). Furthermore, the fluctuating availability of gaming consoles can also be considered as a temporally semi-open event, i.e., a reoccurring salient social context that becomes more scam-inducing when stocks are low.

Ultimately, the combination of a sought-after product and severe issues in its production establishes the social context within which scammers are able to successfully operate. Recognising the emergence of similar circumstances is an important aspect of digital literacy and fraud avoidance. Understanding how the presence of demand and a lack of supply (Kikerpill and Siibak 2021b) provide opportunities for fraud, in particular in online venues where the environment is easily further manipulated (Kikerpill 2021a), is a general skill requirement for staying safe in online environments. Whether the object of desire is some product, service, or even just content, e.g., free streaming of popular TV series or movies, the scam rules are broadly the same: the presence of demand can always be satisfied with pretend supply. With respect to the current example of gaming consoles, the primary aim is to acknowledge the different ways in which the overall cultural importance of certain items and activities intensified the acuteness of an already existing unavailability of products. Social context, in this sense, seems to act as a strengthening agent for underlying wishes and desires. Unlike the commercial holidays example, which incentivises people on the basis of temporal restrictions, the microchip shortage and subsequent gaming console scams placed the focus on objects of desire the demand for which comes and goes as social trends shift.

Discussion and Conclusion

The main objective of the previously presented examples was to explore and explain the ways in which events and social circumstances, i.e., salient social contexts, are or can be used in the dissemination of credible-sounding or looking online scams. In the temporally and geographically restricted comparative tax seasons example, the complexity of the tax system itself and the ease with which people can file their taxes played an important role in terms of the extent of a relevant scam ecosystem (see Kikerpill 2021a). The "criminal calendars" are fixed to the date ranges in which taxes are prepared, filed and returns received. Where the tax filing is made simple for citizens (see e-Estonia 2021), scams appear scarcer as there are fewer points in the process into which scammers can interject their bogus offers. However, when the preparation of taxes is complex enough so as to require the provision of relevant services, scammers will find ways of exploiting this weakness in the system, including how people handle their personal and financial information in the process

(Rafter 2022). Referring to the social context element of the *mazephishing* framework (Kikerpill and Siibak 2021a), the event itself, e.g., the upcoming or ongoing tax season, decreases scammers' workload, because it already provides a seemingly credible reason for contacting people. Hence, knowledge of such processes, including who might be expected to contact a person in these circumstances, is important for fraud avoidance in cases of crime-as-communication.

In the case of Amazon Prime Day and Black Friday sales events, scams follow a well-trodden path of promised gains and a type of fear-of-missing-out experience (Kikerpill and Siibak 2021b), i.e., not buying a product during the sales event means a person would have to wait for the next one. What was particularly important with respect to advancing digital literacy in the area of fraud avoidance, is the fact of how scams follow the well-known social context even if the particular date of the temporally restricted event is changed due to exceptional circumstances (see Johnston 2022). Furthermore, as the observance of such events extends to new areas and communities, the social context that enables respective scams extends along with it. In other words, the more culturally shared (or open) an event is, the larger the geographical range for the dissemination of increasingly believable scams. The application of the first element of the *mazephishing* framework, i.e., the social context element, is relatively easy with well-established "hallmark holidays". Even so, future research could inquire whether this also holds true for culturally restricted events or circumstances, e.g., local fairs that are organised regularly, or other celebrated dates that involve a local commercial element.

The microchip shortage, and the unavailability of popular gaming consoles that the shortage has entailed, shows that a combination of cultural and commercial elements can emerge to enable widespread scams. While these scams are temporally semi-open, i.e., the circumstances that underlie the scams fluctuate, they are concurrently the category that requires the most attention in future research. In comparison with fixed-date (or date range) events such as tax seasons or "hallmark holidays", fluctuating social contexts may be the most difficult to predict in terms of salience for online scams, because current trends in objects of desire or necessity can change quickly and be very different in different parts of the world. Put another way, while we are beginning to learn more about the importance of social context in the dissemination of scams (Carter 2015; Kikerpill and Siibak 2021a; Kikerpill 2021a, Steinmetz et al. 2021), we still lack sufficient information as to what exactly causes some events or circumstances to become salient enough so as to enable the circulation of scams reliant on said context. The gaming console scams are a lone example of how the combination of different circumstances can make for a scam-inducing environment, but more information is required about other similar combinations, i.e., temporal, cultural and geographical aspects that comprise a basis for social circumstances conducive to circulating scams. Given that not all events and circumstances are equally important for members of different communities, and such circumstances are also lived and experienced differently, an important future effort in digital literacy and fraud avoidance must come from employing local knowledge to detect, record and report how salient social contexts create opportunities for scammers. In fact, adopting the *mazephishing* framework for classroom instruction may facilitate this process on a local level in different types of digital literacy courses. By looking at scams that are already detected, it opens the possibility for a more in-depth scrutiny of their timing (e.g., set dates or all-year-round circulation), content and context (e.g., the themes and references to events used in the scams), as well as the channels used for spreading them (e.g., emails, text messages, social media, or bogus websites created for the specific purpose). As it is incredibly difficult to uniformly determine what different people might consider as desirable or necessary under varying circumstances, digital literacy in fraud avoidance is key to mitigating the myriad of crime (as communication) threats that are circulating now or will be circulated in the future.

References

Atkins B, Huang W (2013) A study of social engineering in online frauds. Open J Soc Sci 1(3):23–32. https://doi.org/10.4236/jss.2013.13004

Bischoff P (2020, November 25) 5,000+ Black Friday and Cyber Monday scam sites registered in November. Comparitech. Retrieved February 27, 2022, from https://www.comparitech.com/blog/vpn-privacy/black-friday-scam-website-research/

Bolster Blog (2021, June 16) Amazon scams up 7X leading up to Prime Day. Retrieved February 27, 2022, from https://bolster.ai/blog/amazon-scams-up-7x-leading-up-to-prime-day/

Borel B (2018, October 1) Clicks, lies and videotape. Scientific American. Retrieved July 22, 2022, from https://www.scientificamerican.com/article/clicks-lies-and-videotape/

Button M, Cross C (2017) Cyber frauds, scams and their victims. Routledge

Carter E (2015) The anatomy of written scam communications: an empirical analysis. Crime Media Cult Int J 11(2):89–103. https://doi.org/10.1177/1741659015572310

Carter E (2021) Distort, extort, deceive and exploit: exploring the inner workings of a romance fraud. Br J Criminol 61(2):283–302. https://doi.org/10.1093/bjc/azaa072

Dumoulin I (2019, November 25) The development of Black Friday as a global sales phenomenon. Diggit Magazine. Retrieved February 27, 2022, from https://www.diggitmagazine.com/papers/black-friday-global-sales-

e-Estonia (2021, March 17) Estonian simplicity, American complexity: a tale of two very different tax systems. Retrieved February 27, 2022, from https://e-estonia.com/a-tale-of-two-very-different-tax-systems/

e-Estonia (2022) e-Tax. Retrieved February 26, 2022, from https://e-estonia.com/solutions/ease_of_doing_business/e-tax/

ETCB (2021, December 14) Eesti elanike maksutahe kasvas kolmandat aastat järjest. Estonian Tax and Customs Board. Retrieved February 27, 2022, from https://www.emta.ee/uudised/eesti-elanike-maksutahe-kasvas-kolmandat-aastat-jarjest

Feder S (2021, October 12) Understanding the global chip shortage, a big crisis involving tiny components. Popular Science. https://www.popsci.com/technology/global-chip-shortage/

Fowler B (2022, February 14) Valentine's Day romance scams: don't fall for them. CNET. Retrieved February 27, 2022, from https://www.cnet.com/tech/services-and-software/valentines-day-romance-scams-dont-fall-for-them/

Frank A (2020, November 19) Why the new PlayStation and Xbox are such a big deal. Vox. Retrieved February 26, 2022, from https://www.vox.com/culture/21551062/playstation-5-xbox-series-x-price-games-release-date-explained-next-gen

Hadnagy C (2018) Social engineering: the science of human hacking. Wiley

Harrington B (2012) The sociology of financial fraud. In: Cetina KK, Preda A (eds) The Oxford handbook of the sociology of finance. Oxford University Press, pp 393–410

Hatfield JM (2018) Social engineering in cybersecurity: the evolution of a concept. Comput Secur 73:102–113. https://doi.org/10.1016/j.cose.2017.10.008

Hong, J. (2012). The state of phishing attacks. Communications of the ACM, 55(1). https://doi.org/10.1145/2063176.2063197

ITRC (2019, July 11) Scored on Amazon Prime Day? Watch now for scams. Identity Theft Resource Center. Retrieved February 26, 2022, from https://www.idtheftcenter.org/post/scored-on-amazon-prime-day-watch-now-for-scams/

Johnston B (2022, January 4) Prime Day 2022: everything you need to know. Retrieved February 26, 2022, from https://www.expertreviews.co.uk/amazon-prime-day

Kamasa J (2021) Microchips: small and demanded. CSS Analyses in Security Policy 295. https://doi.org/10.3929/ethz-b-000517399

Kantar Emor (2020) Riigiportaali eesti.ee kasutaja rahulolu analüüs: koondaruanne. Retrieved February 27, 2022, from https://www.ria.ee/sites/default/files/kantar_emor_riigiportaali_eesti.ee_rahuloluanaluus_koondaruanne.pdf

Kessler G (2013, April 13) Claims about the cost and time it takes to file taxes. The Washington Post. https://www.washingtonpost.com/blogs/fact-checker/post/claims-about-the-cost-and-time-it-takes-to-file-taxes/2013/04/13/858a97fc-a455-11e2-9c03-6952ff305f35_blog.html

Khonji M, Iraqi Y, Jones A (2013) Phishing detection: a literature survey. IEEE Commun Surv Tutor 15(4):2091–2121

Kikerpill K (2021a) Crime-as-communication: detecting diagnostically useful information from the content and context of social engineering attacks. University of Tartu Press

Kikerpill K (2021b) The individual's role in cybercrime prevention: internal spheres of protection and our ability to safeguard them. Kybernetes 50(4):1015–1026. https://doi.org/10.1108/K-06-2020-0335

Kikerpill K, Siibak A (2019) Living in a spamster's paradise: deceit and threats in phishing emails. Masaryk Univ J Law Technol 13(1):45–63. https://doi.org/10.5817/MUJLT2019-1-3

Kikerpill K, Siibak A (2021a) Mazephishing: the COVID-19 pandemic as credible social context for social engineering attacks. Trames J Humanit Soc Sci 25(4):371–393. https://doi.org/10.3176/tr.2021.4.01

Kikerpill K, Siibak A (2021b) Abusing the COVID-19 pan(de)mic: a perfect storm for online scams. In: Pollock JC, Kovach DA (eds) COVID-19 in international media: global pandemic perspectives. Routledge. https://doi.org/10.4324/9781003181705-25

Klimburg-Witjes N, Wentland A (2021) Hacking humans? Social engineering and the construction of the "deficient user" in cybersecurity discourses. Sci Technol Hum Values 46(6):1316–1339. https://doi.org/10.1177/0162243921992844

Lawson P, Pearson CJ, Crowson A, Mayhorn CB (2020) Email phishing and signal detection: how persuasion principles and personality influence response patterns and accuracy. Appl Ergon 86:103084. https://doi.org/10.1016/j.apergo.2020.103084

Maggi R (2014) Toward a semiotics of digital places. In: Resmini A (ed) Reframing information architecture. Human–computer interaction series. Springer, pp 85–102. https://doi.org/10.1007/978-3-319-06492-5_7

Marcos CM (2021, November 26) How Black Friday got its name. The New York Times. https://www.nytimes.com/2021/11/26/business/how-black-friday-got-its-name.html

Montañez R, Golob E, Xu S (2020) Human Cognition Through the Lens of Social Engineering Cyberattacks. Front Psychol 11:1755. https://doi.org/10.3389/fpsyg.2020.01755

Muriel D, Crawford G (2018) Video games as culture: considering the role and importance of video games in contemporary society. Routledge

Naidoo R (2020) A multi-level influence model of COVID-19 themed cybercrime. Eur J Inf Syst 29(3):306–321. https://doi.org/10.1080/0960085X.2020.1771222

Norris G, Brookes A, Dowell D (2019) The psychology of internet fraud victimisation: a systematic review. J Police Crim Psychol 34:231–245. https://doi.org/10.1007/s11896-019-09334-5

Osborne H (2021, November 25) Black Friday: how to avoid scams when shopping for deals. The Guardian. Retrieved February 27, 2022, from https://www.theguardian.com/money/2021/nov/25/black-friday-scams-deals-save-money-tips

Pew Research (2015, April 10) 5 facts on how Americans view taxes. Pew Research Center. Retrieved February 25, 2022, from https://www.pewresearch.org/fact-tank/2015/04/10/5-facts-on-how-americans-view-taxes/

Pihlak A (2017, March 12) Petturid saadavad maksuameti nimel õngitsuskirju. Õhtuleht. https://www.ohtuleht.ee/792799/petturid-saadavad-maksuameti-nimel-ongitsuskirju

Posick C (2018) The development of criminological thought: context, theory and policy. Routledge

Powell S (2020, May 29) PlayStation 5: sony confident coronavirus won't change release plans. BBC News. Retrieved February 26, 2022, from https://www.bbc.com/news/newsbeat-52851506

Proofpoint (2019) Human Factor Report 2019. Proofpoint, Inc

PurpleSec (2021) 2021 Cyber security statistics: the ultimate list of stats, data & trends. Retrieved February 27, 2022, from https://purplesec.us/resources/cyber-security-statistics/

Raamatupidaja (2016, January 21) Liikvel on tuludeklaratsiooni petukirjad. Retrieved February 26, 2022, from https://www.raamatupidaja.ee/uudised/2016/01/21/liikvel-on-tuludeklaratsiooni-petukirjad

Rafter D (2022, January 26) 5 IRS scams to watch out for this tax season. Lifelock. Retrieved February 25, 2022, from https://www.lifelock.com/learn/identity-theft-resources/irs-tax-scams-to-watch-out-for

Rapp J (2016, January 31) Rahalubaduse varjus peitub pettus. Lõuna-Eesti Postimees. https://lounapostimees.postimees.ee/3075893/rahalubaduse-varjus-peitub-pettus

Rigotti E, Rocci A (2006) Towards a definition of communication context. Foundations of an interdisciplinary approach to communication. Stud Commun Sci 6(2):155–180

Simons H (2014) Case study research: in-depth understanding in context. In: Leavy P (ed) The Oxford handbook of qualitative research. Oxford University Press, pp 455–470

Smith C (2020, November 25) PS5 stock: sony offers hope of more consoles before end of 2020. Retrieved February 27, 2022, from https://www.trustedreviews.com/news/ps5-stock-sony-offers-hope-of-more-consoles-before-end-of-2020-4110995

Smith D, Aguilar N (2021, November 21) Don't fall for these clever Black Friday scams this year. Retrieved February 27, 2022, from https://www.cnet.com/tech/services-and-software/dont-fall-for-these-clever-black-friday-scams-this-year/

Snowdon D, Churchill EF, Munro AJ (2001) Collaborative virtual environments: digital spaces and places for CSCW: an introduction. In: Churchill EF, Snowdon DN, Munro AJ (eds) Collaborative virtual environments: digital places and spaces for interaction. Springer, pp 3–20

Sobak K (2014, December 30) Järjekordne petuskeem üritab maksu- ja tolliameti nimel raha välja petta. ERR. Retrieved February 25, 2022, from https://www.err.ee/527151/jarjekordne-petuskeem-uritab-maksu-ja-tolliameti-nimel-raha-valja-petta

Steinmetz K, Pimentel A, Goe WR (2021) Performing social engineering: a qualitative study of information security deceptions. Comput Hum Behav 124:106930. https://doi.org/10.1016/j.chb.2021.106930

Taodang D, Gundur RV (2022) How frauds in times of crisis target people. Vict Offenders. https://doi.org/10.1080/15564886.2022.2043968

Tompor S (2021, June 15) Amazon scammers are slick, good at what they do: here's what to watch for. Detroit Free Press. https://eu.freep.com/story/money/personal-finance/susan-tompor/2021/06/15/amazons-scammers-good-what-they-do-heres-what-watch/7633895002/

Torgler B, Schneider F (2007) What shapes attitudes toward paying taxes? Evidence from multicultural European countries. Soc Sci Q 88(2):443–470. https://doi.org/10.1111/j.1540-6237.2007.00466.x

Vargo D, Zhu L, Benwell B, Yan Z (2021) Digital technology use during COVID-19 pandemic: a rapid review. Hum Behav Emerg Technol 3(1):13–24. https://doi.org/10.1002/hbe2.242

Verma R, Crane D, Gnawalli O (2018) Phishing during and after disaster: hurricane Harvey. Resilience Week (RWS):88–94. https://doi.org/10.1109/RWEEK.2018.8473509

Walter D (2019, January 25) Card number, CVV, expiry date are not knowledge elements – or maybe they are? OsborneClarke. Retrieved February 26, 2022, from https://www.osborneclarke-fintech.com/2019/01/25/card-number-cvv-expiry-date-are-not-knowledge-elements-or-maybe-they-are/

Whitney L (2021, June 17) Amazon Prime Day scams resurface for 2021. TechRepublic. Retrieved February 25, 2022, from https://www.techrepublic.com/article/amazon-prime-day-scams-resurface-for-2021/

Work in Estonia (2022, February) Taxes in Estonia. Retrieved February 27, 2022, from https://www.workinestonia.com/working-in-estonia/taxes/

Chapter 5
How Southeast Asia Can Better Arrange and Deliver Internet Policies So as to Defy the Digital Divide

Jason Hung

Introduction

Asia's consumption and economic development have been outpacing most of the world, where half of the top 10 rising brands per the Global Fortune 500 list in 2021 were from the region (PineBridge Investments 2021). While Asia, including Southeast Asia (SEA), continues to rapidly develop, wealth gaps regionwide have been expanding and becoming pronounced (PineBridge Investments 2021). In SEA, some 40% of the labour force still primarily engaged in agriculture, compared to merely 19% who occupy jobs in the services industry. Here poorer SEA countries, including Vietnam and Thailand, share a higher percentage of the agricultural workforce, whereas the countries housing bigger, more urbanised labour markets – Singapore, Malaysia, the Philippines, and Indonesia – enjoy a more services-based labour force (World Bank 2019).

SEA has been accelerating urbanisation and initiating digitalisation in recent years, where regions, led by Singapore, Jakarta, Manila, Bangkok, Ho Chi Minh City, and Kuala Lumpur, have been incorporating technologies in their development to build "smart cities." According to a report published by McKinsey Global Institute (2018), digital urbanisation in SEA may be worth a total of US$26 billion, hinting at the flood of financial opportunities digital technologies can bring to the region in the long term. In his book *The Digital Economy: Promise and Peril in the Age of Networked Intelligence*, Tapscott (1995) coined digital economies as the range of economic activities that depend on digital computing technologies as a focal point of

The original version of the chapter has been revised. A correction to this chapter can be found at https://doi.org/10.1007/978-3-031-30808-6_14

J. Hung (✉)
Department of Sociology, The University of Cambridge, Cambridge, England, UK
e-mail: ysh26@cam.ac.uk

D. Radovanović (ed.), *Digital Literacy and Inclusion*,
https://doi.org/10.1007/978-3-031-30808-6_5

production. He mentioned building digital economies is human-centric, relying heavily on policymakers to formulate policies and regulations and strengthen the protection of intellectual property rights, technological specialists to establish the complex Internet system, and educators to enhance relevant human capital (comprising skills and knowledge) learnt by Internet users (Dolores and Spath 2020). Internet users can exploit the digitalisation of economies to acquire opportunities in career and educational development and social interaction (i.e., communicate with families and friends), so long as they gain sufficient access to e-services (Sharma and Grote 2019).

While rapid digitalisation creates benefits for SEA countries and populations, economic gaps are widened under such a process. This is because some SEA countries have more limited digital technological access and coverage than others, restricting their economic gains (Clavier and Ghesquiere 2021). Digital maturity within the region is divided into four tiers by country. Singapore leads the region in digitalisation as a country known as an advanced global innovation pioneer. The country has successfully delivered digital responses to the outbreak of the coronavirus (COVID-19), by, for example, supporting their public health interventions and offering online platforms for private consumption. Brunei, Malaysia, and Thailand belong to the second-tier category, where each of these three has also satisfactorily utilised technological assets to develop their digital economies and respond to the COVID-19 pandemic. For example, digital transaction, instead of traditional payment with the use of coins or notes, has been widely used in Thailand, especially during the pandemic epoch. Vietnam, Indonesia, and the Philippines are grouped as the third-tier digital economies. They have introduced some digital solutions but have yet to fully incorporate such digital assets into the public health responses or advancement of local populations' everyday life. Cambodia, Laos, and Myanmar, being the fourth-tier digital economies, have an unsatisfactory degree of digital penetration and literacy, rising the alarm that they will further be economically left behind shall they continue to fail to initiate digitalisation (ibid).

The regional digital pioneer, Singapore, introduced the smart city initiative, a programme that originated from its Smart Nation Vision, as early as 2014 to seek to harness data, networks, and information and communications technology (ICT). COVID-19 has accelerated the importance of digital connectivity in every social, economic, and otherwise aspect, leading SEA countries to invest significantly in digital infrastructure as a prominent response to the continuation, if not the growth, of the pandemic (Ingram 2020). As digitalisation is strongly tied to economic growth and public health regulations, the examination of how SEA faces digital exclusion or inclusion, alongside how the region arranges and delivers digital policies to defy the adverse impacts of the digital divide, adds significant value to scholarly debate. The digital divide implies that socioeconomically disadvantaged cohorts are most digitally excluded. These cohorts primarily feature low-income households, the elderly, rural populations, and those who are illiterate. Also, on the enterprise level, the digital divide refers to the situation where small and medium-sized enterprises (SMEs) are at greater risk than larger companies (Kaur et al. 2021). The divide encompasses several dimensions, including digital connectivity and accessibility. More specifically, societies are encountering three levels of the digital divide: the first level in Internet access, the second level in digital literacies and competencies, and the third level in the divide in life opportunities and benefits acquired from the first two stages (Radovanovic et al. 2020). When a rising amount of information and economic opportunities are only available to those who have internet access, those who are digitally excluded and disconnected face impotence to take part in and benefit from the advantages offered by the modern, digitalised economy (Dolores and Spath 2020).

This chapter conceptually explores the digital growth of SEA and provides a systematic overview and analysis of how countries digitally harness and maximise the benefits of online platforms. Then, we discuss how the digital divide, marginalisation, and exclusion as processes, continue to be ingrained in the region, urging a need for intra- and inter-countries' inclusive responses to e-development. Last, we raise some concerns over digital transformation and explore how different SEA parties can respond to these risks. Here responses refer to the provision of policy analysis and recommendations.

Methodology

The systematic review was carried out by searching through the bibliographic databases Google Scholar, Web of Science, and iDiscover (an internal bibliographic database designated for students or staff affiliated with the University of Cambridge). The searched words were ("digital divide," OR "digital exclusion," OR "digital inclusion") & "policies" & ("Southeast Asia" OR "SEA"). Papers published from 1st January 2014 to 31st December 2021 were exclusively included. There were a total number of 209 results displayed. Initial screening was implemented where the following few types of displayed papers were excluded: (i) articles not being written in English, (ii) articles belonging to master's or doctoral theses, and (iii) articles being duplicated. The second-stage screening was applied in which articles (a) rarely addressing policies in relation to digital exclusion/inclusion and (b) barely examining the contexts of SEA were excluded. As a result, a sum of 14 papers were included in this systematic review (see Diagram 5.1). Digital inclusion is known as "the empowerment of individuals and societies to effectively use ICT, enabling them to contribute to and benefit from today's digitalised economies and societies" (Kaur et al. 2021).

Selected literature cover regional digitalisation before and after the outbreak of the pandemic, enabling us to address how the COVID-19 situation has affected the growth of digitalisation and altered any digital policies or responses. While digitalisation has always existed since ICT was introduced, we exclusively studied literature published from 2014 to 2021, since 2014, as mentioned, marked the opening of a national-scale digitalisation initiative in the regional and global leading digital economy – Singapore. Here we analysed and summarised a variety of literature, including policy papers and news articles.

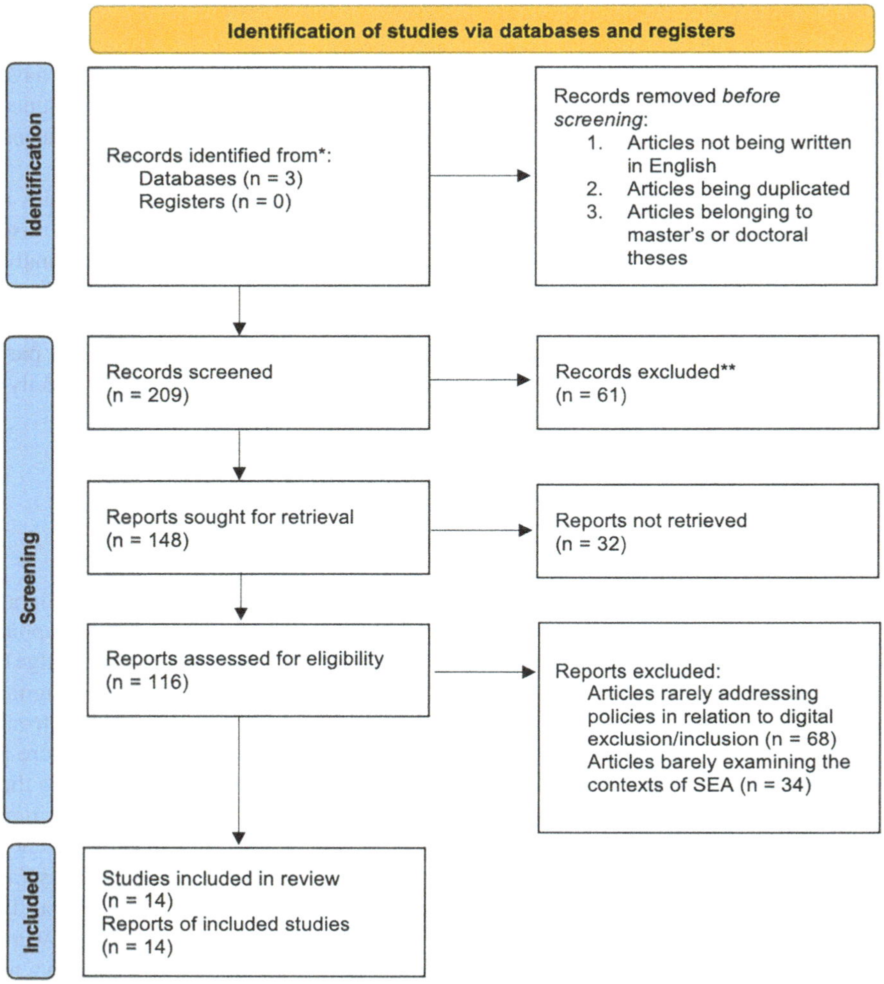

Diagram 5.1 PRISMA 2020 flow diagram for new systematic reviews which included searches of databases and registers only

Studies included in this systematic review were published book(s), policy report(s), policy brief(s), working paper(s), issue paper(s), and newspaper article(s). Given the publication date restriction, all included papers address the digital divide, exclusion, inclusion, and policies in a timely fashion. All retrievable displayed results from the three databases were downloaded and reorganised. We screened all papers accessible for eligibility check and decided to include 14 papers where the contents primarily or partially satisfy the eligibility check.

Findings

As SEA has been undergoing rapid digitalisation, it is important to justify such a decision by discussing the benefits nationwide or regionwide digitalisation offers. Therefore, we will, first, briefly present digital growth within the region and how economic or otherwise opportunities can be generated. Second, we will problematise how digital development renders certain populations to be left behind, triggering a digital divide, marginalisation, and exclusion. In response to these challenges, third, we will engage in the discourse on how SEA countries have responded to the digital problems by forming policies to enhance the degree of digital inclusion and growth. In the following examination, a total of 10 SEA countries, namely Singapore, Indonesia, Thailand, Cambodia, Laos, Myanmar, the Philippines, Brunei, Malaysia, and Vietnam, will be involved in the scholarly discussion. Again, in terms of digital development, Singapore belongs to the first-tier; Thailand, Brunei and Malaysia fall into the second-tier; Indonesia, the Philippines and Vietnam are categorised as the third-tier; and Cambodia, Laos and Myanmar occupy the fourth-tier.

Digital Growth

In SEA, digitalisation results in several major benefits, including economic gain and recovery. COVID-19 has increasingly popularised the use of Singapore's Grab and Indonesia's Gojek, two umbrella mobile applications that offer a marketplace of delivery services and e-offerings, regionwide (Dolores and Spath 2020). When social distancing rules apply, these umbrella digital applications have allowed SEA populations to order delivery food, make contactless payments, and do online shopping. To mitigate the economic harms of the pandemic, the use of these mobile applications has facilitated SEA populations to undertake peer-to-peer donations, where many have performed good deeds by sending delivery products to their beloved or even strangers who are seemingly in need (Clavier and Ghesquiere 2021).

Followed by Singapore's smart city initiative in 2014, Indonesia, as the largest SEA economy (by GDP) that houses the fourth largest population on the globe, launched its 2020 Go Digital Vision campaign in 2015 to introduce and grow its digital economy. Not only have local technological start-ups emerged, digitalisation has also benefited disadvantaged populations and businesses within the country. For example, local e-commerce platforms for agriculture and fishers have been set up; eight million SMEs have been digitalised; and a broadband network accessed by over 150 municipalities nationwide has been built (Dolores and Spath 2020). Digitalisation has therefore helped Indonesian populations and businesses open up economic opportunities, and mitigate the financial risks posed by the pandemic.

Thailand is another SEA country that is desperate to enhance its digital innovation in order to boost its economic growth. Bangkok launched the Thailand 4.0 Innovation-Driven Economy strategy in 2020, aiming to transform the country into a developed nation by advancing its infrastructure and connectivity. Part of the strategy is to consolidate and expand its digital economy by issuing more visas to global talents and companies that share the skills and assets of building smart cities and raising the digital literacy of its own population (Dolores and Spath 2020).

While the West is introducing de-urbanisation and remote education and work in the post-pandemic epoch, SEA is experiencing a growth of urbanisation where waves of an influx of youths to first-tier cities have been seen. This phenomenon is caused by SEA youths' hopes to acquire more economic opportunities which are dominated by the most advanced regional cities. As SEA youths are centralised in most urbanised cities, this prompts the local governments to arrange and deliver digital policies, by, for example, advancing broadband networks and teaching citizens digital skills in big cities, in a more convenient, centralised manner (PineBridge Investments 2021).

Digital technology upsets traditional means of learning, business, communications, and governing. Other than Singapore, Indonesia, and Thailand, other neighbouring countries should speed up their development of digital technology and drive productivity, raise innovation, encourage economic growth, and boost labour productivity in response to how COVID-19 has disrupted our traditional approaches to living (Ingram 2020). Not only do SEA countries need digital growth, but they also need the realisation of digital inclusion too. By enhancing citizens' digital literacy, the competitiveness and wages of employees will raise, and, from a macro-perspective, countries can increase their share of global trade. Digital inclusion is necessary to build an equitable, sustainable digital society, ensuring that disadvantaged populations will not be left behind and become a liability that heavily requires the government's financial support (Kaur et al. 2021). However, to date, an entrenched, expanding digital divide between those with and without digital assets has been seen in SEA. This is detrimental to the economic sustainability within and between SEA countries. It is, therefore, necessary to discuss, in this chapter, what public and private actors should do to advocate for the equalisation and inclusion of digital assets among local populations (Kaur et al. 2021; Sharma and Grote 2019).

So far, in SEA, some 150 million adults, comprising 31% of the regional population, are digitally excluded (Kaur et al. 2021). While this shows that the digital divide is ingrained within the region, the data also hint that SEA countries have significant potential for economic gains shall they be able to promote digital inclusion. An estimated US$150 billion in revenues can be earned from digital economy-relevant activities in the region per year (World Bank 2019). If digital inclusion is fully incorporated, it is explicit that the gain in revenues will rocket. However, as per the World Economic Forum, the pandemic has worsened the digital divide globally where, to date, 55% of the world's population are digitally unconnected (Sridhar 2021). By bridging the digital divide, at least US$15 billion is anticipated to be unlocked each year in SEA (Kaur et al. 2021). These contexts and statistics show that it is necessary to address the growing problem of the digital divide in the post-pandemic epoch, in order to boost economic gains regionwide.

One of the digital divide concerns between SEA countries is digital finance. In Cambodia, Laos, and Myanmar, less than 30% of adults have a digital bank account.

In Myanmar particularly, only one-sixth of bank account holders access their accounts digitally. However, in Malaysia and Thailand, a majority of adults have access to a digital bank account (World Bank 2019). By expanding access to digital payments – such as person-to-person mobile payments, investment, and remittances – SEA populations' digital connectivity can be enhanced. Digital financial inclusion allows informal sectors to be incorporated into the formal economy and avoids individuals being exploited by traditional moneylenders and unresponsive conventional bankers. To date, 198 million of the 400 million SEA population are unbanked (Ingram 2020). This demonstrates regional economies, including Cambodia, Laos, and Myanmar, should accelerate their development of digital finance, especially when there is a growing belief that the pandemic would linger around for a prolonged duration.

The Philippines is a highlighted example that has offered digital finance amid the pandemic. Manila has partnered with United States Agency for International Development (USAID) during the COVID-19 epoch to develop ReliefAgad, a digital application to allocate pandemic relief funds to beneficiaries' bank accounts or e-wallets. Since the pandemic hit, the Philippines has demonstrated its willingness to shift financial services and redistribution to e-platforms (Lee and Lingad 2021). In doing so, the Philippines can offer contactless financial redistribution in a more convenient and swifter manner, which can minimise Manila's administration expenses. In fact, using mobile applications to streamline the distribution of financial support and aid to poorer populations is not a novel approach in SEA. However, the employment of mobile applications for such purposes has by far been accelerated regionwide amid the COVID-19 era (Clavier and Ghesquiere 2021).

Aside from the efforts delivered by public actors, their private counterparts have also maximised the benefits of digital finance to promote financial inclusion. For example, the online payment platform offered by Grab – known as GrabPay – has enabled customers who have an absence of physical bank accounts to make purchases. Unbanked borrowers are given the opportunities to use alternative data, including e-commerce transactions, to prove their creditworthiness. PayLater, a service offered by Grab to allow their loyal users to pay for goods on credit, is a relevant pioneering example. SEA commercial banks, including Thailand's Kasikorn Bank, have also partnered with Grab to offer e-wallets for unbanked individuals. Other commercial banks, such as Indonesia's Jago, have alternatively partnered with Gojek to provide e-wallets for those who are unbanked (Lee and Lingad 2021).

In addition to digital finance, another area demonstrating SEA's digital growth is its Internet usage rate. SEA has the highest rate of Internet usage on the globe. On average, SEA's Internet users spend some 3.6 h per day on the e-platform. Thai users have the highest usage rate among SEA countries, spending 4.2 h each day on the Internet, followed by Indonesian users at 3.9 h daily. SEA's Internet usage rates are by far higher than those of users from dominant Western Internet markets (note: American users spend 2 h per day on the Internet while British cohorts spend 1.8 h every day) (World Bank 2019). These statistics show that more SEA populations have incorporated the use of the Internet into their everyday life. If the majority of them use the Internet for economically and socially desirable purposes, such as building social connectivity, operating e-businesses, and studying digitally, a growth of digitalisation can significantly raise their productivity, wealth, and reserves of human capital.

To build virtual social connectivity, a mainstream approach nowadays is to use social media more frequently, especially when in-person gatherings may be barred during the pandemic. The social media usage rate in SEA is among the highest on the globe. Here a range of social media and telecommunications applications, led by LINE, WeChat, and Facebook and its ownership of WhatsApp and Instagram, have been widely in use in SEA (World Bank 2019). Social media, along with friends, families, and web search engines, is the primary source of information when customers are making purchasing decisions. SEA consumers get information related to products and services on social media, and they are inclined to share brand-related details with others via social media channels (World Bank 2019). Such consumption patterns prompt the expansion of e-commerce opportunities, encouraging more SEA populations and businesses to buy and sell digitally. When selling goods on social media, the suppliers can easily pay delivery fees to Grab or Gojek drivers to deliver the products to the buyers' addresses. The popularisation of the use of delivery applications further facilitates the growth of digital marketisation and consumption in SEA.

SEA's e-commerce platforms reached some US$50 billion (in terms of market size) in 2017 and have the potential to exceed US$200 billion by 2025. Online travel is the dominant category of e-commerce regionwide, with US$26.6 billion led by the growth in hotel and airline online bookings. E-commerce sales of first-hand products, additionally, reached some US$ 10.9 billion (in terms of gross merchandise value) in 2017 from US$ 5.5 billion in 2015 (World Bank 2019). These statistics unveil the strong ties between revenues and digital consumption, in which the e-commerce market in SEA has been significantly and growingly lucrative. Regional leading e-commerce giants, such as Lazada, Shopee, and Tokopedia, have been able to provide readily-accessible, scalable platforms; and smaller retailers have been reaching consumers within and beyond SEA. In 2017 alone, a total of some 200 million digital consumers were seen in the region, resulting in a rise of 50% from 2016 (World Bank 2019). Given the outbreak of the pandemic that has rendered more digital activities, e-commerce transactions should have further surged in the past 2 years. The forecast of the number of Internet users in SEA shows that some 554.48 million individuals within the region will be using the Internet in 2025, compared to only 122.54 million in 2010 (see Graph 5.1). This reflects the noticeable potential of digital growth in SEA.

A key factor of the growth in e-commercing regionwide is the rise of Internet users. In 2011, there were only 127 million Internet users in SEA. At the end of 2017, the number reached 390 million. Over half of the SEA population was already online at the end of 2017, mostly coming from Indonesia, Malaysia, and the Philippines (World Bank 2019). SEA is among the most rapidly developing markets on the globe in terms of Internet penetration (with a compound annual growth rate (CAGR) of 13% from 2011 to 2016, whereas the world's average was at some 8%).

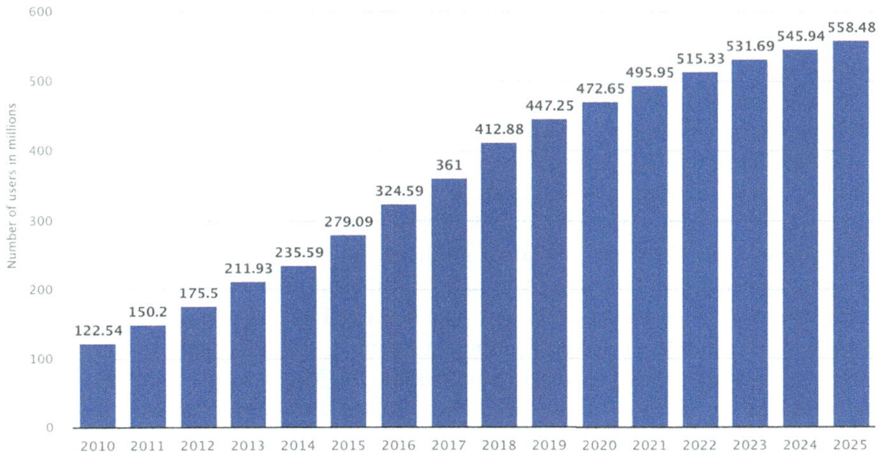

Graph 5.1 Forecast of the number of Internet users in SEA from 2010 to 2025 (in millions)

Indonesia, Thailand, and the Philippines recorded a CAGR of 16%, 15%, and 14%, respectively, during the six-year span (World Bank 2019). If the digital divide can be addressed and the rate of digital inclusion improves, a tremendous rise in digital activities, including those related to e-commerce, will be seen in SEA.

Digital Divide, Marginalisation, and Exclusion

Populations, especially those earning lower incomes, the elderly, village-based cohorts, and the less educationally attached, that are excluded from digital activities are deprived of technological empowerment. These populations have a proclivity to face financial exploitation through paying higher transaction fees and encountering more exposure to intermediaries. They heavily rely on cash and experience substantial constraints from accessing credit, prompting them to be trapped in a vicious cycle of poverty (Sridhar 2021).

It is therefore essential for SEA populations to develop digital literacy through, for example, formal education. Since the outbreak of the pandemic, schools and universities regionwide have been suspended. In-person classes have been shifted to e-learning in order to avoid COVID-19 transmission. The sudden switch to e-teaching has exposed SEA to a flood of functional concerns because a large proportion of the regional population has an absence of Internet access (Jalli 2020). For example, most universities in Indonesia have switched to the e-learning mode. However, many local schools and students have failed to deliver and enjoy digital resources, respectively, engendering students to self-study at home (Jalli 2020). Therefore, not only do students of less advantaged backgrounds enjoy limited digital access, but less funded schools and universities may lack the financial resources to develop e-teaching platforms. Educational digitalisation in such contexts faces double barriers.

In SEA, as of June 2021, only Brunei, Malaysia, Singapore, Thailand, and the Philippines each shared over 80% of Internet penetration. Vietnam, Indonesia, and Cambodia, alternatively, enjoyed more than 70% of Internet penetration. On the contrary, Laos and Myanmar had just above 50% of Internet penetration (see Graph 5.2). A large segment of the SEA population fails to afford unlimited, stable Internet connections. Even for those who are digitally connected, Internet speeds offered by the same Internet providers may vary across different regions – where populations living in most urbanised cities enjoy a much faster Internet speed than those from villages. For example, In Kuala Lumpur, residents can enjoy Internet speed at 800 megabytes per second. However, in Sarawak in East Malaysia, the Internet is either not accessible or at a very slow speed (Jalli 2020). These unveiled how the digital divide is spatially driven. Individuals in East Malaysia are worried that the poor Internet connection would not suffice to facilitate e-learning, e-commercing, and remote work (Jalli 2020). Such digital exclusion significantly impacts residents' everyday life in an adverse manner.

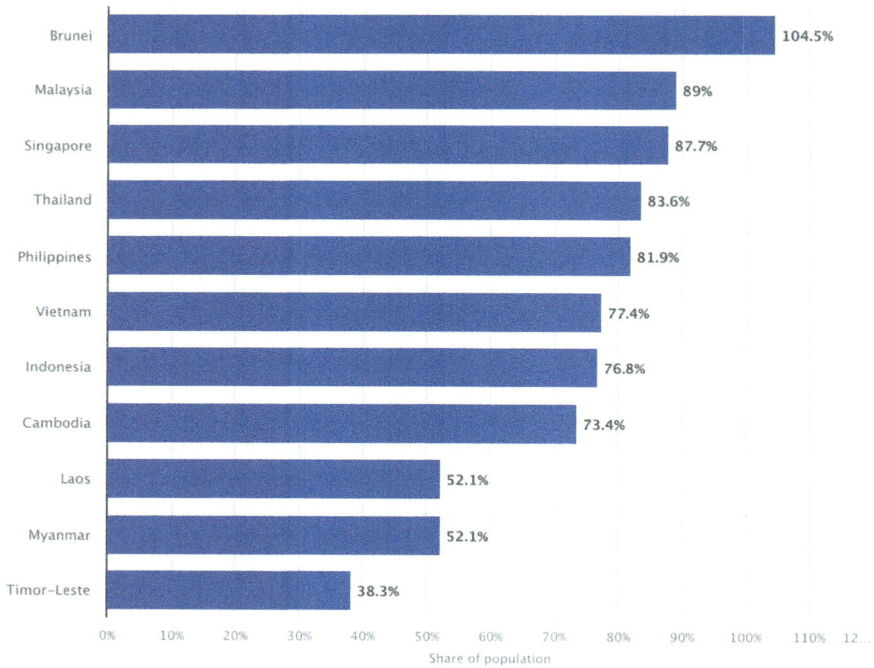

Graph 5.2 Internet penetration in SEA as of June 2021, by country. (Source: Statista 2022)

Digital Policies

Digital inclusion, alternatively, plays a role in social equalisation. This means by providing digital inclusion can beneficiaries enjoy more equal access to education, choices of consumption, and free flow of information. Digitally included cohorts can capitalise on the wealth of online resources to expand their levels of human capital acquisition, facilitating their ability for upward social mobility. Digital inclusion, therefore, bridges the gap between those of different sociodemographic and socioeconomic backgrounds (Kaur et al. 2021). In addition, digital inclusion functions as a governance stabiliser. By promoting digital inclusion, coverage of citizens, civic outreach, and access to public services would raise, resulting in an administration of better governance. The provision of e-government services helps deliver better efficiency and reduce administration costs for different governmental departments (Kaur et al. 2021).

Digital Inclusion

Throughout the globe, including in SEA, Internet access is tied to more challenges than Internet coverage. As of 2015, 79% of Asia was covered by the 3G network. Yet, just more than half of the Asian population consumed the Internet. Here usage rates varied across SEA. Some 70% of the Vietnamese population had Internet access, relative to between 40% and 50% of the Cambodian, Indonesian, and Philippine populations. Laos, Myanmar, and Timor-Leste had even more unsatisfactorily low Internet access rates, each below 32% (Ingram 2020). To enhance Internet access, given many SEA populations fail to afford to use the Internet, Wholesale Open Access Network (WOAN) should be widely built. To date, over 160 regulatory authorities throughout the globe have established some degree of open Internet access infrastructure (Ingram 2020). Those who cannot afford privately purchased Internet networks can gain Internet access provided in public areas. As a result, SEA should raise the coverage of the WOAN provision, otherwise, the benefits of such infrastructure would be limited.

The lack of market competition is another entrenched challenge SEA is facing. An estimated 589 million individuals reside in countries lacking competition which keeps Internet prices high. More than 260 million individuals only have one choice of major mobile network operator to choose from, and one or two operators usually dominate(s) most SEA countries. The limited competition in different levels of the broadband value chain significantly affects the affordability and quality of mobile networks provided. A report published by Alliance for Affordable Internet (2019) states that the transition from an Internet market with little to no competition to a market with robust competition can save Internet users up to US$3.42 per gigabyte (Ingram 2020). Regulatory authorities in SEA should actively invite more potential mobile network operators to submit proposals requesting for entering the market. This helps raise healthy competition and ensures Internet users within the region can benefit from more affordable and higher-quality mobile networks. Facilitating competitive pricing can improve more financially friendly digital inclusion. A pro-competition regulatory framework incentivising innovation and investments can ensure broadband offerings are more competitive. This can additionally help improve the quality, availability, and stability of rural networks which are often a less attractive market to private, commercial investors (Kaur et al. 2021).

Digital Infrastructure

To enhance digital inclusion, SEA countries are required to invest more in the development of digital infrastructure, especially in villages, towns, and underserved cities where digital connectivity is less unaffordable or unavailable. Building broadband infrastructure nationally is a must. Also, in archipelagic countries, such as Indonesia and the Philippines, new, sustainable, and innovative technologies, including floating solar-powered networking rigs, should be further built (Kaur et al. 2021; World Bank 2019). Not only should broadband infrastructure be universally built, but the construction of high-speed broadband facilities is particularly preferable. This is because the entitlement to high-speed Internet services is more user-friendly, and the operation of local businesses would be significantly hampered if high-speed Internet cannot be accessed. It is noteworthy that low- and middle-income developing countries in SEA all enjoy speeds over mobile and fixed broadband at a lower level than the Organisation for Economic Co-operation and Development (OECD)'s average (World Bank 2019). Such contexts imply that digital marginalisation and exclusion remain significant in lower-income regional economies, while their richer counterparts offer desirable, high-speed Internet services primarily. The entrenched digital divide will plausibly widen the wealth gaps between regional economies unless digital infrastructure can be widely delivered in poorer nations.

However, fixed high-speed broadband infrastructure is costly to be built and poorer countries face financial challenges to fund such construction projects. In Malaysia, ranked 42nd out of 182 countries by International Telecommunications Union, fixed broadband cost was 1.1% of GNI per capita in 2015. Vietnam ranked at 66th with 1.8%; Thailand ranked as low as 98th with 3.9%; and the Philippines and Indonesia even ranked 126th and 133rd with 7.53% and 9.51%, respectively (World Bank 2019). As fixed rather than mobile (i.e., a much cheaper alternative) broadband infrastructure is required for data-intensive business operations (World Bank 2019), public and private sectors should collaboratively fund projects to build more fixed, high-speed broadband facilities. Divisions of the Association of Southeast Asian Nations (ASEAN) and Asian Development Bank (ADB) focusing on building urbanised, smart cities should prioritise distributing more financial resources to lower-income SEA countries to build such infrastructure. Governments' implementation of regulatory reforms is needed to raise competition at every level of the broadband value chain (both wholesale and retail) for the purpose of optimising investments in underserved or even unserved regions (World Bank 2019). Such an intervention is essential to ensure more stable and reliable Internet connectivity, making SEA cities smarter, more sustainable, and more economically competitive.

Currently, innovative technologies are incorporated into the construction of digital infrastructure. Smart grids, integrated water management systems, and intelligent transport systems are some of the broadband-enabled technologies that are adopted to enhance efficiencies when developing a more inclusive digital economy (WEF 2016). With the exploitation of innovative technologies, SEA countries are endeavouring to advance broadband connectivity regionwide, which serves as an instrument for online trade and e-commerce (WEF 2016). Therefore, SEA's economic competitiveness can improve; and poorer countries that have now constructed more fixed broadband infrastructure enjoy a higher degree of financial sustainability in the long term.

Digital Education: Digital Literacy

While SEA endeavours to improve digital infrastructure, populations within the region need to, simultaneously, develop the literacy and skills to use the Internet. Digital education should be included in the national curriculums as soon as possible; and technical or professional training should be redesigned to focus on raising the workforce's digital skills (Kaur et al. 2021). These skills range from basic computer usage (including the use of Microsoft and email system) to advanced skills (such as data analytics and coding), as well as alternative soft skills like online communications and collaboration (World Bank 2019).

By 2030, six (i.e., 12 million) to eight (i.e., 17 million) percent of the SEA non-agricultural workforce will be displaced by technology. Education should therefore focus more on developing the next generations' digital literacy and skills and English proficiency, in order to cultivate more analytical talents (Woetzel et al. 2014). Otherwise, those who lack sufficient digital literacy will be left behind by the urban labour markets, and they may plausibly be trapped by social immobility and poverty.

When redesigning and delivering education or training on digital literacy and skills within the region, governments should ensure they do not overlook the importance of including the 14% of SEA populations who remain living under the international poverty line (Tan 2020). Only by improving poor children's and adolescents' digital literacy as well as working adults' digital skills, can they have an opportunity for urban labour market entry in the digitalisation era. Otherwise, they will further be excluded from the urban economies, and have to take up agricultural jobs that may significantly be displaced by technology.

To date, there is a lack, if not an absence, of published policy frameworks on digital (il)literacy in SEA individually or collectively. This is an evident policy gap in which individual SEA governments, alongside the ASEAN, should endeavour to develop and deliver clear guidelines on how policies in relation to enhancing populations' digital literacy are or will be formed. The ASEAN should also provide guidance to member states on designing digital literacy-relevant policies that accommodate the needs of their own citizens but also the interests of SEA as a whole.

Digital Finance

Despite the growth of digitalisation in SEA, the World Bank Global Findex data report that as low as 19% of financial account holders regionwide access their accounts digitally. The figure is much lower than the average of the globe's middle-income countries at 27%. The current system of digital finance excludes a significant proportion of the regional populations from capitalising on FinTech tools to improve the convenience of their everyday life and their economic well-being. Governments need to expand their digital coverage, facilitating the use of digital payments to distribute pensions, government services, or cash handouts. Government-run authorities should also universalise the use of digital ID schemes to help citizens open bank accounts in an easier manner. Such a policy especially benefits the unbanked populations who may not have a qualified legal document to open a physical bank account (World Bank 2019). Regionally, the ASEAN Digital Integration Framework and the ASEAN e-Commerce Agreement facilitate member states to deliver the full potential of digital integration. In support by the Framework and Agreement, to date, five member states (namely Singapore, Malaysia, Indonesia, Thailand, and Brunei) have completely digitalised their identity systems, and another three member states (namely Vietnam, Cambodia, and Laos) are piloting their digitalised foundational ID system (UNCTAD 2020).

Governments employing digital finance can also support local business operations by, for example, offering online licensing and permit approvals (World Bank 2019). Here e-government services can minimise administration costs and time, incentivising more potential local business owners to start their own businesses regionwide. Especially when digital finance allows Internet users to pay for goods and services remotely and instantly, more potential business owners will be encouraged to set up their e-commerce shops (World Bank 2019).

However, if SEA does not develop local populations' digital literacy and skills, individuals may not understand how to get access to or use digital finance. Therefore, the development of digital literacy and finance should be undergone simultaneously, in order to ensure that Internet users know how to maximise the use of e-financing services to receive payments or start businesses. Otherwise, digitally illiterate cohorts will face substantial digital exclusion, lacking opportunities to instrumentalise online tools, and services to enhance their own economic well-being.

Discussion

Aside from the ingrained digital divide and some forms of digital barriers SEA is encountering, there are other concerns over digital transformation that regulatory authorities within the region need to address thoroughly to build a safe, reliable digital economy. The public is concerned about the growing and substantial adoption of financial services provided by Grab and Gojek. While such an act is facilitating financial growth and inclusion, data breaches reported over the years have raised questions that financial regulations should address. Authorities should regulate these digital-financial services to an extent that balances consumer protection and the integrity of national financial systems (Lee and Lingad 2021).

As digitalisation has been speeding up regionwide since the outbreak of COVID-19 where a growing number of SEA citizens rely on digital services, it is urgent that regulatory authorities should strengthen their relevant legal framework as soon as possible. If regulatory authorities do not respond timely, further misuse of data, unchecked surveillance, and among other digital threats will jeopardise the credibility and stability of the use of digital services (Ingram 2020, 2021).

While digitalisation can be instrumentalised to alleviate poverty and improve national economic growth in the long term, it can also heighten political divisions and exacerbate financial inequality (Ingram 2020). To date, less than half of SEA countries have detailed data protection laws; and data protection authorities' capacity remains significantly limited. Policies are usually coordinated locally but not regionally, rendering it difficulties for businesses and individuals to understand what regulations apply whenever their data move across borders within SEA (World Bank 2019). Therefore, ASEAN and ADB specialists should work with regulatory authorities from every SEA country to collaboratively implement data protection laws.

Currently, only Singapore, Malaysia, the Philippines, and Thailand enjoy comprehensive data protection statutes (note: Singapore passed its *Personal Data Protection Act 2012*; Malaysia enforced its *Personal Data Protection Act 2010*; the Philippines enacted its *Data Privacy Act 2012*; Thailand issued its *Personal Data Protection Act B.E. 2562* (Assi 2013; Cohen et al. n.d.)). As cross-border e-commerce services have been expanding within SEA, there is a strong incentive to advance cooperation in formulating and passing data protection laws collectively and regionally.

Conclusion

Not only should SEA continue to build smart cities, but governmental authorities and private actors should also collaboratively ensure they establish digital platforms for studies, professional work, finance, consumption, and other purposes without financially dividing those who are socioeconomically privileged and disadvantaged.

Public-private initiatives can be developed in each of the SEA countries, where the public sector has the leverage of better familiarity with digital administration while the private sector delivers the establishment of digital infrastructure and facilities in a rather efficient, cost-effective, and better-quality fashion. Each regional country should accelerate its digital development to mitigate any economic challenges posed by the pandemic while ensuring most, if not all, citizens have rather equal access to digital resources. They should also actively build more reliable and open-accessible digital infrastructure, to ensure disadvantaged cohorts would not be left behind from the digital growth regionally. Government authorities should also collaboratively develop national and regional regulations for the purpose of facilitating more individuals and businesses to turn their everyday activities digital.

Digitalisation in SEA is like a goose that lays a golden egg. It creates a tremendous lucrative financial market and economic growth for each involved country, so long as digital development and inclusion are maturely built. However, digitalisation is also a double-edged sword. While a raft of financial resources can be generated, it worsens the financial and digital exclusion suffered by digital (semi-) illiterate, less-educated, and economically underprivileged individuals. Therefore, only by achieving an equitable, reliable, stable, and safe digital environment can SEA populations and businesses improve their financial well-being in a sustainable manner.

References

Alliance for Affordable Internet (2019) 2019 Affordability report: lack of competition in broadband markets keeping millions offline. Retrieved from https://a4ai.org/news/2019-affordability-report-lack-of-competition-in-broadband-markets-keeping-millions-offline/. Accessed 6 Feb 2022

Assi G (2013) South East Asia: data protection update. Bryan Cave, Singapore

Clavier F, Ghesquiere F (2021) Leveraging digital solutions to fight COVID-19: lessons from ASEAN countries, Research & policy briefs no. 41. Washington, DC: World Bank

Cohen J, Santaniello D, Oo N, Chitranukroh A, Bui T (n.d.) Regional guide to cybersecurity and data protection in mainland Southeast Asia. Tilleke & Gibbins. Washington, DC

Dolores M, Spath K (2020) Women and the future of the digital economy in Asia: decent work for all? The Friedrich-Ebert-Stiftung. Washington, DC

Ingram G (2020) Development in Southeast Asia: opportunities for donor collaboration. Chapter 2: the digital world. Centre for Sustainable Development at Brookings. Washington, DC

Ingram G (2021) Bridging the global digital divide: a platform to advance digital development in low- and middle-income countries, Brookings global working paper no. 157. Centre for Sustainable Development at Brookings. Washington, DC

Jalli N (2020) Lack of Internet access in Southeast Asia poses challenges for students to study online amid COVID-19 pandemic. The Conversation, 17th March. Retrieved from https://theconversation.com/lack-of-internet-access-in-southeast-asia-poses-challenges-for-students-to-study-online-amid-covid-19-pandemic-133787. Accessed 13 Sept 2022

Kaur S, Low J, Dujacquier D (2021) Bridging the digital divide: improving digital inclusion in Southeast Asia. Roland Berger. Washington, DC

Lee K, Lingad D (2021) Digital growth and financial inclusion in Southeast Asia. Centre for Strategic and International Studies. Retrieved from https://www.csis.org/blogs/new-perspectives-asia/digital-growth-and-financial-inclusion-southeast-asia. Accessed 13 Sept 2022

Mckinsey Global institute (2018) Smart cities in Southeast Asia, Discussion paper July 2018. Mckinsey & Company. Washington, DC

PineBridge Investments (2021) Age of Asia: rise of a multipolar world. Retrieved from https://www.pinebridge.com/en/investment-opportunities/age-of-asia. Accessed 13 Sept 2022

Radovanovic D, Holst C, Belur B, Srivastave R, Houngbonon G, Quentrec E, Miliza J, Winkler A, Noll J (2020) Digital literacy key performance indicators for sustainable development. Social Inclusion 8(2):151–167

Sharma R, Grote U (2019) Determinants of internet use among migrants in South-East Asia: a case study of internal migrants in Thailand and Vietnam, no. 58. International Organisation for Migration. Washington, DC

Sridhar R (2021) Bridging the digital divide is key to building financial inclusion. Forbes Business Development Council. Retrieved from https://www.forbes.com/sites/forbesbusinessdevelopmentcouncil/2021/09/10/bridging-the-digital-divide-is-key-to-building-financial-inclusion/. Accessed 13 Sept 2022

Tan A (2020) Tech groups must address the digital divide in South-East Asia. Financial Times, 9th January. Retrieved from https://www.ft.com/content/f5818706-3093-11ea-a329-0bcf87a328f2. Accessed 13 Sept 2022

Tapscott D (1995) The digital economy: promise and peril in the age of networked intelligence. McGraw-Hill. Washington, DC

United Nations Conference on Trade and Development (UNCTAD) (2020) Strengthening knowledge and skills through innovative approaches for sustainable economic development. The United Nations. Washington, DC

Woetzel J, Tonby O, Thompson F, Burtt P, Lee G (2014) Southeast Asia at the crossroads: three paths to prosperity. McKinsey Global Institute. Washington, DC

World Bank (2019) The digital economy in Southeast Asia: strengthening the foundations for future growth. World Bank. Retrieved from https://openknowledge.worldbank.org/handle/10986/31803. Accessed 13 Sept 2022

World Economic Forum (WEF) (2016) State of ICT in Asia and the Pacific 2016: uncovering the widening broadband divide. United Nations ESCAP. Washington, DC

Part II
Digital Literacy Textures and Education

Chapter 6
The Digital Divide and Higher Education

Kerry Russo and Nicholas Emtage

Introduction

As higher education moves to blended learning environments, a digital divide is emerging in the Australian higher education sector. This divide is predicated on differing digital skills and usage patterns, not access to digital devices. In turn, many students transitioning to university do not have the necessary digital skills required to participate in a digital setting.

Is the use of learning technologies contributing to inequity in higher education, an inequity due to differing digital experiences, digital resources, and usage patterns? COVID shone a spotlight on this inequity that is the digital divide. The move to remote learning saw an expanse of this divide sometimes referred to as digital poverty. Students lacking digital skills, access, and devices were further disadvantaged during remote learning (Bashir et al. 2021; Pentaris et al. 2021; Summers et al. 2021). If the digital divide is to be overcome, universities cannot continue to assume the digital fluency of commencing students.

Using a quantitative approach, the chapter provides an analysis of the digital divide in Australian higher education, examining how differing digital fluency stages influence perceived preparedness for university study. The chapter conceptualises the growing inequalities arising from a widening digital divide, by investigating impacts on the student experience, digital fluency, and secondary schooling digital opportunities. Reporting on the research question: "What is the relationship

---

This chapter is dedicated to the memory of our dear friend, colleague and mentor Lynne Eagle.

K. Russo (✉) · N. Emtage
James Cook University, Bebegu Yumba Campus, Townsville, Australia
e-mail: Kerry.russo@jcu.edu.au; nicholas.emtage@jcu.edu.au; https://www.jcu.edu.au; https://www.jcu.edu.au

D. Radovanović (ed.), *Digital Literacy and Inclusion*,
https://doi.org/10.1007/978-3-031-30808-6_6

between socioeconomic, sociocultural/geographic indicators and the digital divide?" empirical data on the digital divide provides an examination to determine the link between digital fluency, socioeconomic status, sociocultural capital, digital identity, and student self-reported preparedness and digital skills. About 409 first-year business students were surveyed at regional and urban Australian universities. See Appendix 6.1 for the study questionnaire.

Our proposition is that digital fluency is predicated on prior digital experiences and that socioeconomic and geographic indicators influence the attainment of digital fluency, influences which subsequently impact perceived preparedness for university study.

Background

The digital divide is defined as a gap in digital knowledge and a gap in opportunity, ability, and efficacy (van Deursen and van Dijk 2011; Warschauer et al. 2010). This digital divide is not based on access to digital devices only. Though inequitable access to digital resources creates a disadvantage (van Deursen and van Dijk 2011), in this study, access was not the primary issue as numerous Australian secondary schools offer a school-issued laptop scheme. This scheme was anticipated to level the playing field for students from disadvantaged backgrounds.

As stated above, the digital divide emanates from different levels of digital fluency. Digital fluency is defined as the ability to use digital technologies to interpret, problem-solve, create, and reformulate knowledge (Wang et al. 2013). Briggs and Makice (2011) define digital fluency as "an ability to reliably achieve desired outcomes through use of digital technologies" (p. 64). In this chapter, digital fluency is defined as the ability to successfully move with ease in a digital environment. In simple terms, digital fluency is to create rather than consume in a digital environment.

Digital fluency is an important skill for a twenty-first-century workforce. In a constantly changing digital environment, university graduates need to be digitally fluent to be competitive in a future workplace. Digital fluency assists in future-proofing graduates and builds resilience for entry to a post-COVID disruptive workforce.

Pre-COVID concerns were being raised about the relationship between digital technologies usage and inequality. An increase in youth disengagement and alienation from formal institutions was noted (Broadbent and Papadopoulos 2013; Caluya et al. 2018). Broadbent and Papadopoulos (2013) announced being part of the digital divide in the twentieth century disconnects you from a part of your world that now exists for others. This disconnect was distinctive during the COVID pivot. As some students struggled with online exam platforms, online lectures, and navigating the digital learning space in overcrowded home environments (Bashir et al. 2021; Pentaris et al. 2021).

The COVID pivot has changed how we deliver education forever. Therefore, the need to deliver digital learning environments which are fair and equitable begins with the digitally fluent student. A review of the literature demonstrates many hurdles to achieving digital fluency. Beginning at secondary schools, if teachers are not provided with access to professional development or technical support staff, they are reluctant to engage with learning technologies (Warschauer et al. 2010). No teacher wants to be in front of a class having technology issues. Caluya et al. (2018) noted a relationship between social economic status (SES) and differences in digital skills and knowledge. Multiple researchers have found digital fluency inequities to be socio-economically driven (OECD 2021; Radovanović et al. 2015; van Dijk 2006; Warschauer et al. 2010). Mominó et al. (2008) and Castaño-Muñoz (2010) established private schools produced students with higher digital fluency even with lower technological resources than their state school counterparts. This led to their contention that high levels of technological resources did not equate to higher digital skills of students but rather schools' ineffective use of the curriculum (Mominó et al. 2008; Castaño-Muñoz 2010).

Methodology

A convenient sampling technique was used for the study which was conducted at a regional Australian university and an urban Australian university across the 2017 and 2018 1st year business student cohort. A total of 259 questionnaires were distributed at the regional university in marketing and management lectures with 236 returned completed: a 91% response rate. At the urban university, 179 questionnaires were handed out in marketing and management lectures with 173 questionnaires returned completed: a 96% response rate. The high response rates could be contributed to the questionnaires being handed out in paper form and collected in the lectures. Students were informed they did not have to participate in the survey and were entitled to hand back a blank questionnaire. A combined total of 409 participants were thus surveyed. The survey instrument was paper based with data then recorded in an SPSS (v.25) data file that was subsequently used for quantitative analyses.

The study participants were surveyed to determine whether disadvantage indicators impact digital fluency and contribute to a digital divide in higher education. Survey questions centred on each respondent's beliefs about the importance, motivation, constraints, and opportunities of technology. The survey questions were based on a 5-point Likert scale. Self-reported digital literacy skills, information fluency, and the respondents' online enrolment experiences were measured to find the level of digital fluency. Based on these indicators, the measurements were assessed against demographic factors and access to digital devices. The definition of who is from a disadvantaged household is from the Australian Bureau of Statistics Socio-Economic Index For Areas which ranks areas in Australia in terms of their relative disadvantage with those households in the bottom 25% of the state classed as 'Low socio-economic status'.

Descriptive analysis was used to examine students' demographic features, digital access, and digital fluency indicators. Pearson's Chi-Square tests for association were performed and a Cramer's V test was executed to assess the strength of association where the testing involved two or more categorical variables and one-way ANOVAs were used in the instances where testing involved examining continuous and categorical variables. The results presented in this article are only in cases where tests indicated significance at the 95% confidence level, that is, p values from the tests were less than 0.05.

Results

Table 6.1 below illustrates the respondents' demographic characteristics and school background.

Survey questions designed to measure access to digital devices during secondary schooling established 51% had a school-issued laptop, 73.9% had a personal computer/laptop, and 92.6% responded they had used computers/digital technologies throughout secondary schooling.

Table 6.1 Demographics of respondents

		Distribution	
Variable	Category	Frequency	Percentage (%)
University	Urban	173	42
	Rural	236	58
Gender	Male	151	43
	Female	202	57
Age group	School leaver (<20 years)	226	65
	Post-school leaver (20–24 years)	114	32
	Mature aged (>24 years)	10	3
Socioeconomic status	High	33	13
	Medium	154	62
	Low	62	25
First in family	First in family	141	41
	Not first in family	207	59
Geographic location	Urban	62	19
	Regional city	145	43
	Rural	60	18
	International	68	20
Secondary school type	Private independent	63	17
	Catholic	91	24
	State (government)	132	35
	International	88	24

Questions relating to access to digital technologies sought to determine whether participants had access to a digital curriculum during their secondary schooling and 53% of respondents had access to a school LMS, which suggests a digital curriculum. The presence of an LMS at a respondent's secondary school was revealed throughout the analyses to be strongly related to the development of digital fluency, more so than a school-issued laptop, socioeconomic or sociocultural status apart from students' parent's use of digital resources at work and home.

Analysis of the survey responses revealed the correlations between the origin of students and disadvantage indicators as well as the presence and quality of school LMS's. These rural and regional participants were more likely to be from medium-low socioeconomic backgrounds, first in family, and have attended a State or Catholic school in comparison to their urban counterparts (Fig. 6.1).

Access to an LMS in secondary school recorded the highest variance in the study against all disadvantage indicators and perceived digital ability. School LMSs contributed to students' perception of preparedness for university study (Fig. 6.2): 89% of respondents with a school LMS agreed to be well prepared for university-level study, whereas 79% of respondents that did not have a school LMS felt they were well prepared.

The chart (Fig. 6.3) illustrates the differences in LMS access across geographic and school categorical variables. Of note is that urban schools were much more likely to have an LMS than regional, or rural schools regardless of school type. Furthermore, Fig. 6.3 illustrates that private schools were more likely to have an LMS than State schools, apart from those in rural areas. Within regional and rural areas State schools were less likely to have an LMS, and regional city schools overall were more likely to have an LMS than rural schools.

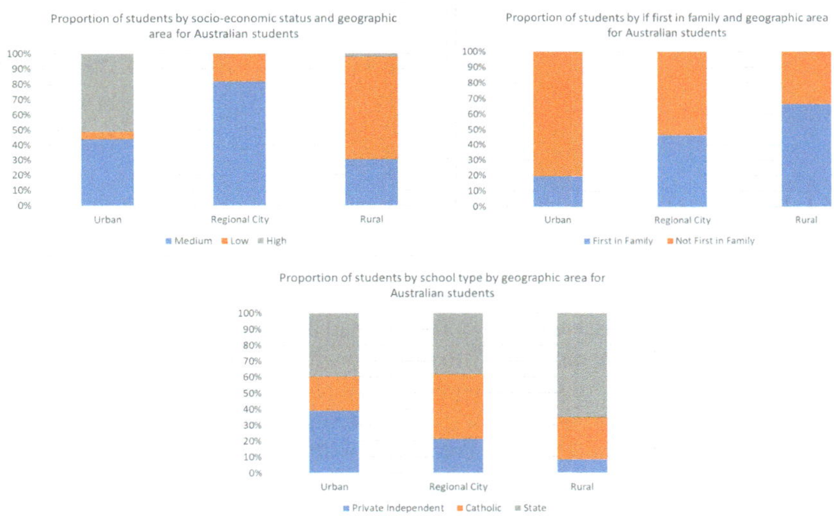

Fig. 6.1 Socio-economic background and if first in family at university by geographic area for Australian students

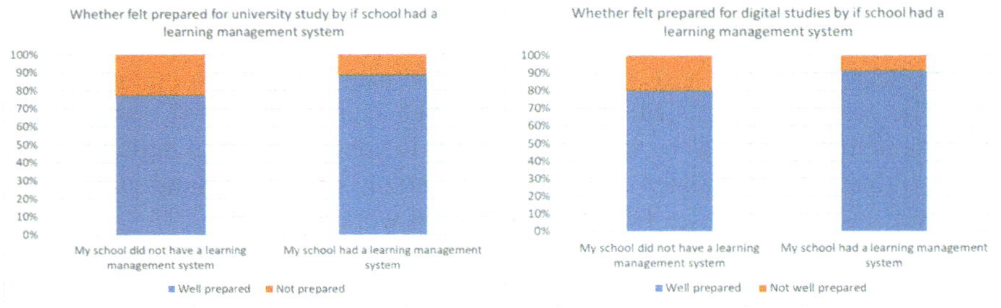

Fig. 6.2 If felt well prepared for university and digital studies by whether attended a school with a Learning Management System

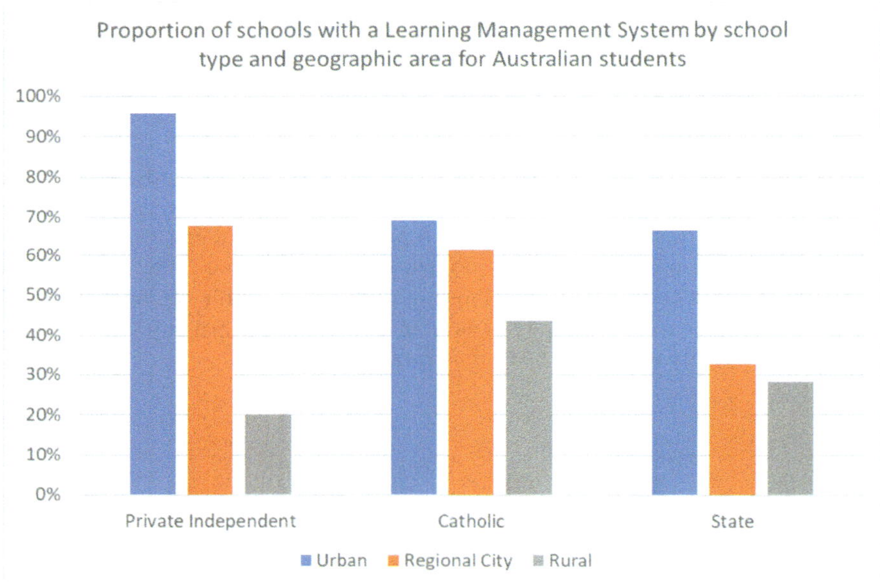

Fig. 6.3 Proportion of schools with a learning management system by school type and geographic area

Participants who required assistance to enrol online ($p < .003$) or contacted the university for enrolment assistance ($p < .015$) or had difficulty setting up their class registration ($p < .001$) were more likely to disagree with preparedness for university. These other indicators of digital preparedness were found to differ significantly based on geographic location, where 68.1% of regional and rural participants required assistance to enrol online compared with 50% of urban participants ($p < .024$). A further 75.8% of regional and rural participants had difficulty setting up their class registration compared with 17% of urban participants ($p < .005$). Furthermore, participants from regional and rural schools consistently rated themselves lower on a scale of 1–5 for digital literacy proficiency than urban school participants.

The presence of an LMS was a critical factor but the impact on their sense of preparedness was mediated by other factors. Multiple disadvantage indicators were related to preparedness e.g. geographic area, school type, SES, and sociocultural factors, as well as the presence of an LMS. An LMS is more likely to be present in an urban private school which in turn is more likely to be populated by non-first in family and higher socioeconomic students. Consequently, while the results indicate access to an LMS in secondary school enhanced students' sense of preparedness for university study, the results do not definitively support an LMS as able to overcome all challenges to developing digital fluency in the presence of the disadvantage indicators.

The relationships between the variable "perceived preparedness by secondary school for university study" and students' demographic and educational backgrounds are of great interest. Respondents who disagreed with the preparedness variable were more likely to be female, from a regional or rural location, have attended a State or Catholic school, be first in the family, not have access to a school LMS, required help to enrol online, and contacted the student centre for enrolment assistance. As observed earlier the 'disadvantage indicators' are correlated and also more present with students from rural and regional areas.

Analysis of socioeconomic status and access to an LMS during secondary schooling illustrates 33% of participants from a low SES background had access to an LMS, compared with 91% of high SES background participants ($p < .001$) (Fig. 6.4). These results were reiterated in students that are first in family at university and the access to an LMS, with 59% of first in family not having access to an LMS ($p < .001$) compared with 37% in not first in family. This narrative continued across all disadvantage indicators including geographic location, with 68% of rural participants not having access to an LMS at secondary school, compared with 21% of their urban counterparts ($p < .001$). A further 60% of State school participants did

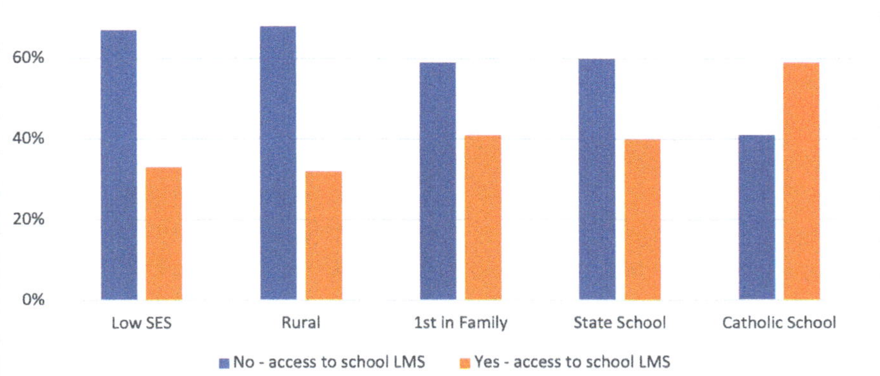

Fig. 6.4 Access to a School LMS with disadvantaged indicators

not have access to an LMS compared with 25% of private independent school participants ($p < .00$).

Geography appears to play a role in attenuating differences in the presence of LMS's in schools observed for students from different socio-cultural backgrounds. For example, while students from low socioeconomic backgrounds generally attended schools that were less likely to have an LMS than students from high socioeconomic backgrounds, students from medium socioeconomic backgrounds were less likely to have attended a school with an LMS if they were in a regional city (54%) than in an urban area (74%), and less likely again if they were in a rural area (40%). The same pattern holds true for students that are first in family (FiF) at university, with 67% of FiF students from urban areas attending schools with an LMS compared to 41% of FiF students from regional cities and 34% of FiF students from rural areas (Figs. 6.5 and 6.6).

So how do all these factors interact in relation to students' preparedness? While the presence of an LMS at their school was related to their sense of preparedness for university and digitally based studies, the Australian students' perception of their preparedness for digital studies was most strongly related first to their parents' degree of use of digital devices in their work and home and their parents keeping up with the latest technologies (combined as a measure 'parental influence digitally') (Fig. 6.7). Only 5% of those who reported their parents had high (strong) use of digital technologies felt unprepared for digital studies at university compared to 20% of those whose parents used digital technologies less. For the second group, the presence of a learning management system at their school appears to help students feel better prepared for digital studies as 10% of those whose school had LMS felt unprepared compared to 27% of those whose school had no LMS. For students whose parents had high digital technology use, the type of school they attended

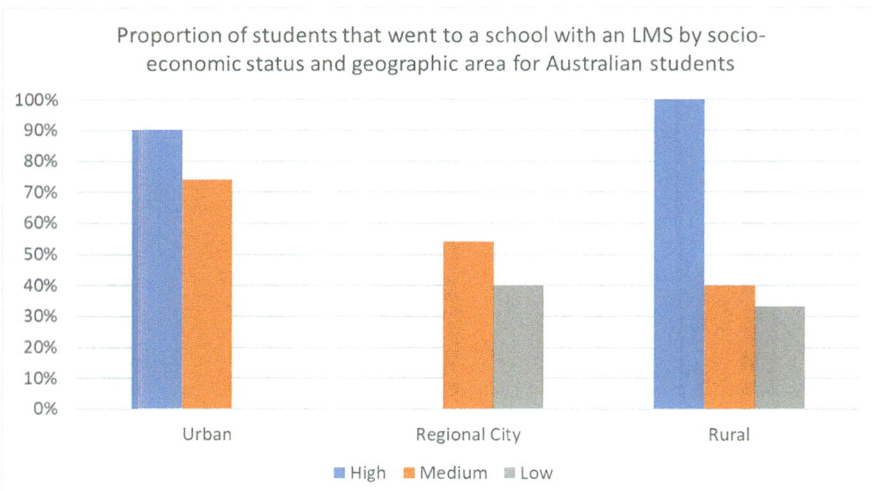

Fig. 6.5 Proportion of Australian students who attended a school with a LMS by geographic area and socioeconomic status

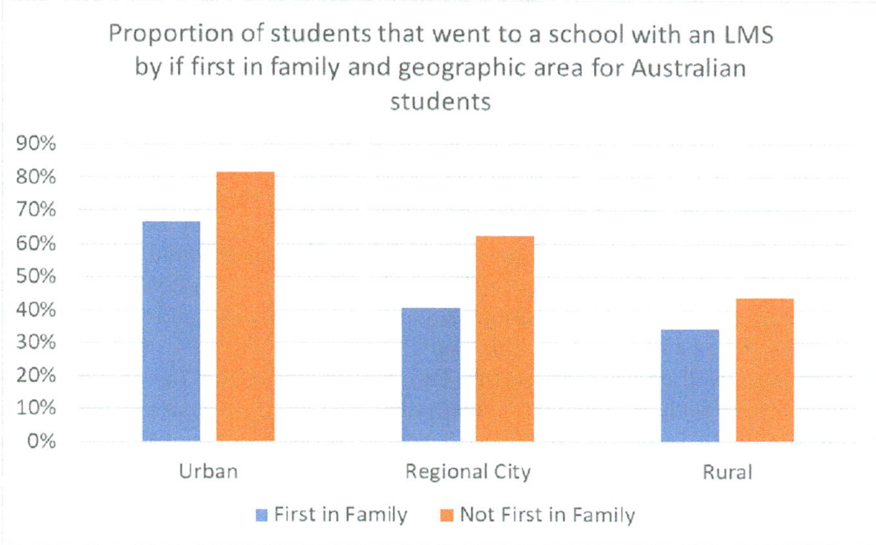

Fig. 6.6 Proportion of Australian students who attended a school with a LMS by geographic area and if first in family at university

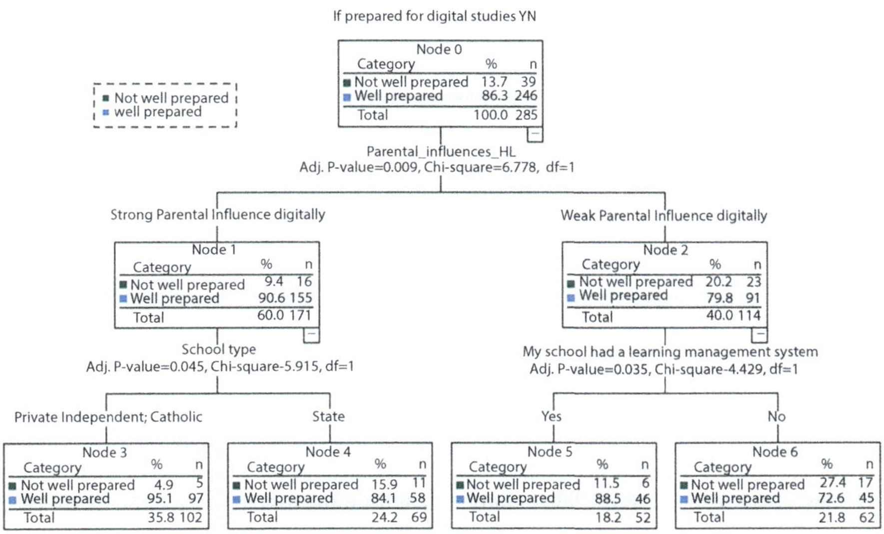

Fig. 6.7 CHAID decision tree of factors related to differences in Australian students perceived preparedness for digital studies

differentiated between them, with 5% of those at private independent or Catholic schools feeling unprepared compared to 15% of those who attended state schools.

No access to a school LMS had a significant impact on digital skills. Proficiency levels were consistently rated lower by participants without access to a school LMS in Outlook Calendar or equivalent, online tests and quizzes, editing video and sound recordings, postings to blogs, forums, and wikis, posting to social networking sites, and uploading videos to social networking sites. These participants were also less likely to critically evaluate information for fairness, validity, and currency.

Digital attitude, mindset, and perceived digital skills were found to be statistically significant against the categorical variables of access to an LMS, geographic location, access to a school-issued laptop, enrolment assistance, and first-in family.

Gender played a role as well, with 81.6% of females and 68.7% of males agreeing they were well prepared ($p < .002$). Participants who scored low in digital fluency indicators such as online enrolment issues also disagreed that they were well prepared ($p < .003$). Lower proficiency levels in Adobe also equated to reports of not being well prepared ($p < .037$). Participants who rated themselves as underprepared consistently rated themselves lower in proficiency in all the digital literacy platforms, for example, Excel, Outlook Calendar, but the results were not statistically significant. This was again evident in questions relating to parental digital skills and self-rating of digital technology skills. Participants who rated themselves and their parents as lacking digital technology skills were also more likely to disagree that their school prepared them well for university-level study.

Discussion

What does this mean? The narrative that has unfolded has illustrated key differences in how school influences, digital experiences and access to digital technologies have influenced study participants' perception of their digital fluency and perceived preparedness for university studies. These digital influences and experiences, when linked to disadvantage indicators such as socioeconomic/sociocultural capital, geographic location, and school type, indicate a relationship between access and application of digital resources and the development of digital fluency.

The findings demonstrate that digital fluency is pronounced in individuals from higher socioeconomic status and sociocultural capital, who attended schools with an LMS, and who had greater access to family or friends who could assist with digital issues. These students were more likely to be digitally fluent and report being well prepared for university-level study These findings that the provision of support builds digital fluency reinforce the work of (Caluya et al. 2018; Devlin and O'Shea 2012; Warschauer and Matuchniak 2010). The need for higher education to create inclusive and supported digital learning environments is clear from the results.

The overall results indicate a digital divide which reflects wider society has emerged in higher education. The design of university digital learning environments assumes students are digitally fluent, especially school leavers who had access to school-issued laptops but these laptops/access to resources in themselves are not sufficient to develop digital fluency.

Digital proficiency and the distribution and application of digital resources appears to be the major contributing factor to the developing digital divide. Research into the link between SES factors, school digital resources, and the digital divide and its impact on Australian higher education is not well-developed as yet; however, the results from this study indicate that digitally underprepared students participating in higher education could be further disadvantaged if unsupported in a digital learning environment.

Students identified as not digitally fluent were more likely to consider not being prepared for university or learning in a digital environment. The research has also shown certain conditions have to be met before digital fluency can be achieved. To use a metaphor, in research on keys to smallholder forestry, Byron (2001) refers to finding a key to unlock the greatest potential gain. Byron (2001) states conditions under which outcomes can be reached are like "a door with many locks", and all locks have to be opened before potential can be realised. Byron's metaphor can be applied to the development of digital fluency in secondary school graduates. In order to unlock the door to digital fluency, four keys are needed. If any of the keys are missing, the secondary school graduate would struggle to achieve digital fluency. The four keys or conditions that have to be met to be digitally fluent are:

1. Access and experience in a digital environment
2. Opportunities to learn in a digital curriculum
3. Experiences in creating, not just consuming, digital knowledge and
4. Constructing a technical and social identity through digital immersion

Therefore, the research has determined that digital fluency was achieved through experience and immersion in a digital environment. Figure 6.8 illustrates the cycle of maintaining digital fluency. Similar to language acquisition, digital fluency requires immersion in a digital environment and practice of digital skills. Therefore, the digitally fluent can move up and down the scale accordingly to their immersion, opportunity to practice, and experience.

The research established that perceived preparedness for university-level study was impacted by commencing students' digital fluency. Therefore, if immersive digital experiences foster digital fluency, the pivot to online learning during COVID should lead to increased levels of digital fluency. However, there is evidence COVID has broadened the digital divide in education. This evidence is captured in the extra services required to support university students during online examinations (Montenegro-rueda et al. 2021).

Figure 6.9 proposes considerations for building digital fluency in commencing university students who may not be digitally prepared to study in a digital learning environment. Building digital fluency in a university student requires an awareness of the student's past experience. Universities have to immerse students in a digital environment which may be a foreign environment for the student therefore supports are required. The university student must be provided with opportunities to practice within a supportive environment. These opportunities help to instill resilience and proficiency and will most likely lead to digital fluency.

The design of many digital learning environments assumes students are digitally fluent. Therefore, the preparation of students to study in a digital learning

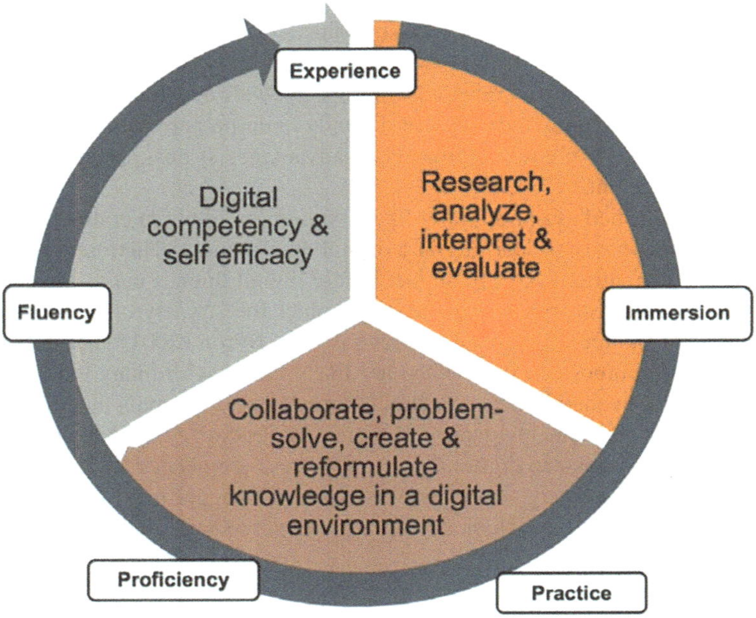

Fig. 6.8 The cycle of digital fluency

Fig. 6.9 Considerations for building digital fluency

environment is paramount. Of greatest importance is the need to orientate, scaffold, and support the digital experience. The higher the level of complexity, the higher support required.

If universities are preparing business students to take their place in an ever-changing digital world, universities need to produce graduates who can create, interpret, and evaluate information, who move with ease in a digital environment to solve problems and create and generate knowledge. Universities need to graduate digitally fluent.

Implications for Theory and Practice

A digital divide based on digital proficiency is present in Australian schools and universities. The study established a relationship between disadvantage indicators and levels of digital fluency. This relationship has clear implications for theory and practice in that digitally resourcing disadvantaged schools does not increase digital fluency. However, digitally resourcing schools combined with clear curriculum direction and teacher professional development in digital pedagogies would likely increase digital fluency and perceived preparedness for university study. If change is not instigated, digitally underprepared students entering higher education could be further disadvantaged and underprepared to study in a digital learning environment.

Limitations and Future Work

The over-reliance on self-reported digital skills is a limitation of the study. Participants may have been likely to rate their digital skills higher than their actual digital skills. Self-reported school identity also contributes to the limitations of the study. The inclusion of secondary school inputs would have strengthened the study. Secondary school interviews and reviews were not included in the studies due to constraints in the research design.

A further limitation is that self-reported digital skills were not linked to academic performance. It would be of great interest to identify whether a lack of digital fluency impacts negatively on academic performance. There are many threads in the research that could not be explored in depth. Further research areas could include:

- A large-scale digital fluency study in Australian higher education.
- Digital fluency impacts on the preparedness of disadvantaged and under-represented students for university study.
- Digital curriculum/LMS implications in secondary schools.
- Building teacher capacity in digital pedagogies in secondary schools.
- Business student academic performance and digital fluency.

Conclusion

The preceding research demonstrates a link between access to a learning management system (LMS) or digital curriculum during secondary school and disadvantage indicators. Access to a school LMS consistently produced higher self-reported digital skills than those without, even when disadvantage indicators were present. The issue of perceived preparedness for university study and/or a digital learning environment was also linked to participants who had access to a school LMS. The results indicate discrepancies in how participants perceived their preparedness for university study and could be indicative of systemic problems with the Australian education system. Rural, regional, low socioeconomic, low socio-cultural capital, and state-school participants were less likely to have had access to a digital curriculum during secondary schooling and were less likely to report preparedness for university study. Conversely, these disadvantage indicators were moderated if participants had access to an LMS or digital curriculum.

The digital divide in higher education is emanating from the distribution, use, and allocation of secondary schooling digital resources and prior experience. The resourcing of secondary schools with school-issued laptops did not increase digital fluency or perceived preparedness for university study in itself. However, the implementation of a digital curriculum or LMS produced significant outcomes in the development of digital fluency. These findings illustrate the influence of digital immersion in the formation of fluency. Resourcing schools without a clear digital curriculum does not increase digital fluency. Schools' level of development of their LMS's also followed a gradient whereby the better-resourced private schools had better-resourced LMS's in terms of equipment, maintenance, and training of the teachers using the system compared to State schools that had LMS's in general. A poorly developed and maintained LMS would offer little benefit to students compared to well-resourced and run LMS's.

The COVID pivot highlighted the need for educational institutions to create supportive digital learning environments. Learning environments that are fair, equitable, and responsive to student needs (Nordmann et al. 2020). Now, more than ever, creating an intentional digital learning experience built on the knowledge of students' digital needs will ensure equity (Bashir et al. 2021). If the digital divide is to be conquered, universities cannot continue to assume the digital fluency of commencing students.

This divide is impacting students' sense of preparedness and their learning experiences. Our proposition that the digital divide is predicated on digital proficiency has been supported by empirical data on the link between digital skills, SES, socio-cultural capital, and self-reported preparedness for university study. Unless effective support structures and curriculum design that build digital fluency are embedded in education, inequity will continue to grow. Further investment is required to build educators' digital skills to facilitate learning environments that promote digital fluency and prepare students for a globally disruptive post-COVID workforce. The OECD first reported a link between the digital divide the internet usage in 2001 (OECD 2001). Twenty-one years later, the time has come to stop talking the talk and begin working towards a fairer educational system that builds digital human capital and levels the playing field. If education is to be transformative, it needs to be supportive and accessible for all.

Acknowledgements The authors would like to acknowledge Lynne Eagle and David R Low. Their valuable guidance, imitable support and insight led to the development of our research.

Appendix 6.1: Questionnaire 6. The Digital Divide and Higher Education

Student number or Login ID:	
1. What secondary school did you attend?	

Please respond to the following questions	Yes	No
2. I had a school issued laptop during my secondary schooling. If yes, please circle if you could: Take home or Use at School only	②	①
3. I had a personal computer or laptop during my secondary schooling	②	①
4. I have used computers/digital technologies throughout my secondary schooling	②	①
5. My school had a Learning Management System eg. Blackboard, Moodle etc	②	①
Think back to when you enrolled in the Bachelor of Business and answer the following questions	**Yes**	**No**
6. It was difficult to enrol online at university	②	①
7. I couldn't enrol online	②	①
8. I needed help to enrol online. If yes, circle who helped you to enrol: Family or Friends or University Staff	②	①
9. I needed to contact the student centre for help to enrol. If yes, circle how you sought assistance: Phone or email or Face to Face	②	①
10. It was difficult to set up my class registrations	②	①

Please indicate how strongly you agree or disagree with the following statements	Strongly agree	Agree	Agree somewhat	Undecided	Disagree	Disagree	Strongly disagree
11. I was well prepared by my school for university level study	⑦	⑥	⑤	④	③	②	①
12. I was well prepared by my school to study in a digital learning environment	⑦	⑥	⑤	④	③	②	①
13. I would rate myself as having excellent digital technology skills	⑦	⑥	⑤	④	③	②	①
14. I grew up using computers/digital technologies	⑦	⑥	⑤	④	③	②	①
15. My parents/caregivers actively use computers/digital technologies in the workplace and home	⑦	⑥	⑤	④	③	②	①
16. My parents/caregivers keep up with the latest trends in technology	⑦	⑥	⑤	④	③	②	①
17. I would rate my parents/caregivers as having good computer/digital technology skills	⑦	⑥	⑤	④	③	②	①
18. I feel it is important to be able to access the Internet any time I want	⑦	⑥	⑤	④	③	②	①
19. I think it is important to keep up with the latest trends in technology	⑦	⑥	⑤	④	③	②	①
20. I believe there is only one right way to use digital technologies	⑦	⑥	⑤	④	③	②	①
21. I can quickly learn how to use new technology	⑦	⑥	⑤	④	③	②	①
22. I am able to jump from one kind of digital technology to another to achieve my goals	⑦	⑥	⑤	④	③	②	①
23. I recognise the potential transformative uses for new digital technologies	⑦	⑥	⑤	④	③	②	①
24. I take comfort with the fact that there is more than one way to use a technology	⑦	⑥	⑤	④	③	②	①
25. I think technologies, not people, always cause success or failure	⑦	⑥	⑤	④	③	②	①
26. I think high social media use always causes a decrease in face-to-face communication	⑦	⑥	⑤	④	③	②	①

27. I often oversimplify or underestimate the role of a new technology	⑦	⑥	⑤	④	③	②	①
28. I understand the types of potential value in using social media	⑦	⑥	⑤	④	③	②	①
29. I have a large number of followers on social media	⑦	⑥	⑤	④	③	②	①
30. I believe change is necessary	⑦	⑥	⑤	④	③	②	①
31. I embrace change as opportunity	⑦	⑥	⑤	④	③	②	①
Please indicate how strongly you agree or disagree with the following statements	Always	Very frequently	Frequently	Sometimes	Rarely	Very rarely	Never
32. I use JCU Library One Search to research my assignments	⑦	⑥	⑤	④	③	②	①
33. I use JCU Lib Guides to research my assignments	⑦	⑥	⑤	④	③	②	①
34. I use Google or other search engines to research my assignments	⑦	⑥	⑤	④	③	②	①
35. I use Google Scholar to research my assignments	⑦	⑥	⑤	④	③	②	①
36. I use Wikipedia to research my assignments	⑦	⑥	⑤	④	③	②	①
37. I only use peer reviewed or academic articles for my assignments	⑦	⑥	⑤	④	③	②	①
38. I use online referencing tools eg. Endnote, Cite this for me or Easy bib	⑦	⑥	⑤	④	③	②	①
39. I critically evaluate information by checking that the content is fair, valid and current	⑦	⑥	⑤	④	③	②	①
40. I evaluate and interpret online sources by checking for bias	⑦	⑥	⑤	④	③	②	①

On a scale of 0 to 5 (with 0 being not competent at all to 5 being expert user) please indicate your competence with the following digital technologies.	5	4	3	2	1	0
41. Microsoft Word or equivalent	⑤	④	③	②	①	⓪
42. Excel	⑤	④	③	②	①	⓪
43. PowerPoint	⑤	④	③	②	①	⓪
44. Email	⑤	④	③	②	①	⓪
45. Outlook calendar or equivalent	⑤	④	③	②	①	⓪
46. LearnJCU	⑤	④	③	②	①	⓪
47. PebblePad	⑤	④	③	②	①	⓪
48. Online Tests eg LearnJCU quizzes, Aplia, Wiley	⑤	④	③	②	①	⓪
49. Posting to Blogs, Forums and Wikis	⑤	④	③	②	①	⓪
50. Creating Blogs, Forums or Wikis	⑤	④	③	②	①	⓪
51. Adobe Acrobat Professional	⑤	④	③	②	①	⓪
52. Graphics packages eg. Adobe Photoshop etc	⑤	④	③	②	①	⓪
53. Post material to social networking sites eg. Facebook, Instagram	⑤	④	③	②	①	⓪
54. Upload videos to social media eg YouTube, Facebook, Instagram, Snapchat	⑤	④	③	②	①	⓪
55. Editing video and sound recordings	⑤	④	③	②	①	⓪
56. Web searches	⑤	④	③	②	①	⓪

References

Bashir A, Bashir S, Rana K, Lambert P, Vernallis A (2021) Post-COVID-19 adaptations; the shifts towards online learning, hybrid course delivery and the implications for biosciences courses in the higher education setting. Front Educ 6. https://doi.org/10.3389/feduc.2021.711619

Briggs C, Makice K (2011) Digital fluency. Building Success in the Digital Age. SociaLens

Broadbent R, Papadopoulos T (2013) Bridging the digital divide – an Australian story. Behav Inform Technol 32(1):4–13. https://doi.org/10.1080/0144929X.2011.572186
Byron R (2001) 16. Keys to smallholder forestry in developing countries in the tropics. In Sustainable farm forestry in the tropics: social and economic analysis and policy. pp 211
Caluya G, Bororica T, Yue A (2018) Culturally and linguistically diverse young people and digital citizenship: a pilot study. Retrieved from Centre for Multicultural Youth, Carlton. http://apo.org.au/node/140616
Castaño-Muñoz J (2010) Digital inequality among university students in developed countries and its relation to academic performance. RUSC Univ Knowl Soc J 7(1). https://doi.org/10.7238/rusc.v7i1.661
Devlin M, O'Shea H (2012) Effective university teaching: views of Australian university students from low socio-economic status backgrounds. Teach High Educ 17(4):385–397. https://doi.org/10.1080/13562517.2011.641006
Mominó JM, Migalés C, Meneses J (2008) L'escola a la societat xarxa. Ariel, Barcelona
Montenegro-rueda M, Luque de la Rosa A, Sánchez-serrano JLS, Fernández-cerero J (2021) Assessment in higher education during the covid 19 pandemic: a systematic review. Sustainability (Basel, Switzerland) 13(19):10509. https://doi.org/10.3390/su131910509
Nordmann E, Horlin C, Hutchison J, Murray J-A, Robson L, Seery MK, MacKay JR (2020) Ten simple rules for supporting a temporary online pivot in higher education, vol 16. Public Library of Science, San Francisco, p e1008242
OECD (2001) Understanding the digital divide. https://doi.org/10.1787/236405667766
OECD (2021) Bridging digital divides in G20 countries. OECD Publishing, Paris. https://doi.org/10.1787/35c1d850-en
Pentaris P, Hanna S, North G (2021) Digital poverty in social work education during COVID-19. Adv Soc Work 20(3):x–xii. https://doi.org/10.18060/24859
Radovanović D, Hogan B, Lalić D (2015) Overcoming digital divides in higher education: digital literacy beyond Facebook. New Media Soc 17(10):1733–1749. https://doi.org/10.1177/1461444815588323
Summers R, Higson H, Moores E (2021) The impact of disadvantage on higher education engagement during different delivery modes: a pre- versus peri-pandemic comparison of learning analytics data. Assess Eval High Educ:1–11. https://doi.org/10.1080/02602938.2021.2024793
van Deursen AJAM, van Dijk JAGM (2011) Internet skills and the digital divide. New Media Soc 13(6):893–911. https://doi.org/10.1177/1461444810386774
van Dijk JAGM (2006) Digital divide research, achievements and shortcomings. Poetics 34(4–5):221–235. https://doi.org/10.1016/j.poetic.2006.05.004
Wang Q, Myers MD, Sundaram D (2013) Digital natives and digital immigrants: towards a model of digital fluency. Bus Inf Syst Eng 5(6):409–419. https://doi.org/10.1007/s12599-013-0296-y
Warschauer M, Matuchniak T (2010) New technology and digital worlds: analyzing evidence of equity in access, use, and outcomes. Rev Res Educ 34(1):179–225. https://doi.org/10.3102/0091732x09349791
Warschauer M, Matuchniak T, Pinkard N, Gadsden V (2010) New technology and digital worlds: analyzing evidence of equity in access, use, and outcomes. Rev Res Educ 34:179–225. https://doi.org/10.2307/40588177

Chapter 7
Students' Use of Social Media and Critical Thinking: The Mediating Effect of Engagement

Asad Abbas, Talia Gonzalez-Cacho, Danica Radovanović, Ahsan Ali, and Guillermina Benavides Rincón

Introduction

Initially, during the time of the COVID-19 pandemic, the continuation of education seemed a difficult undertaking with the use of formal methods of teaching and, as a result, most of the students and their studies suffered due to the transformation of the learning process (Bhardwaj et al. 2021). This rapid transformation in traditional modes of learning forced educational institutions to establish Information and Communication Technology (ICT)-based educational policies (Haq et al. 2021). The adoption of new policies and the implementation of a new education system require basic digital literacy related to the knowledge and skills of available tools (Balasooriya et al. 2018). Digital literacy can also be achieved through informal learning such as social media. Radovanovic et al. (2015) stated that digital literacy

A. Abbas (✉)
Writing Lab, Institute for the Future of Education, Tecnológico de Monterrey, Monterrey, Nuevo Leon, Mexico

School of Government and Public Transformation, Tecnológico de Monterrey, San Pedro Garza García, Nuevo Leon, Mexico
e-mail: asad.abbas@tec.mx

T. Gonzalez-Cacho
School of Architecture, Art, and Design, Tecnológico de Monterrey, Monterrey, Nuevo Leon, Mexico

D. Radovanović
Department of Technology Systems, University of Oslo, Oslo, Norway

A. Ali
School of Economics and Management, Zhejiang Sci-Tech University, Hangzhou, China

G. B. Rincón
School of Government and Public Transformation, Tecnológico de Monterrey, San Pedro Garza García, Nuevo Leon, Mexico

D. Radovanović (ed.), *Digital Literacy and Inclusion*,
https://doi.org/10.1007/978-3-031-30808-6_7

can also be achieved through social media and for younger generations Facebook was their introduction to the Internet.

Kaeophanuek and others (2018) state that the use of social media can enhance digital literacy through the development of three competencies: (1) information skills (searching, interpreting, evaluating, and synthesizing information), (2) digital tools usage (ability to use software applications, manage personal information on networks and ethics), and (3) digital transformation (producing new forms of information, creating new knowledge, and producing digital innovation). For the deployment of these competencies, students must use critical thinking. Moon and Bai (2020) analyze the use of social media for promoting political and civic engagement in youth and concluded that although time spent on social media does not guarantee engagement, designing social media-based activities in an online learning context promotes engagement, and as a consequence, the development of critical thinking skills by developing media literacy. In digital contexts, critical thinking is needed to filter the flux of digital information (Moon and Bai 2020).

Dabbagh and Kitsantas (2012) define social media as a broad concept that makes reference to a variety of network tools or technologies that focus on the social use of the Internet for communication, collaboration, and creative expression. Social media include can be used for resource sharing, the creation of collaborative workspaces, social networking sites, and collaborative work using web-based office tools (Dabbagh and Kitsantas 2012). This research focuses on the students' use of Facebook, Instagram, Twitter, LinkedIn, and YouTube. Recent trends in published literature provide evidence related to the rapidly increasing use of social media in society (Malik et al. 2020). Social media tools can be used to develop students' soft skills and also can have an effect on current education systems and teaching methods (Al-Rahmi et al. 2015; Mafarja et al. 2022). At the same time, social media allows students to continue their learning processes and also enables opportunities for them to engage with their peers to collaborate within or outside of the institution (Clark et al. 2017). The literature review showed that the informal use of technology for pedagogical purposes can encourage students to engage in interdisciplinary course activities (Bedenlier et al. 2020).

The impact of technology on learning within different disciplines has been reviewed in several studies (Radovanović et al. 2020; Le et al. 2022). The results suggest that students prefer technology for learning because it makes the learning experience more pleasant or enjoyable (Radovanović et al. 2020; Le et al. 2022; Ahmad et al. 2021; Armstrong and Georgas 2006). Further, the studies also show that technology in learning improves attitudes towards learning such as motivation, engagement, and commitment (Ahmadi 2018). Nevertheless, a non-formal study has been conducted to review the degree to which engagement might mediate the relationship between social media and critical thinking among students of higher education (Sherman 2013). To explore the social media-based activities and engagement of students to virtually interact with their peers and share useful information or share knowledge through social media platforms which help them to improve their soft skills (Gonzalez-Cacho and Abbas 2022). To achieve the objectives of this empirical research, we conducted an online survey among undergraduate students from Architecture and Civil engineering programs and statistically analyzed collected data using the Jamovi R-based software application. The purpose of our research is to understand the mediating role of engagement between the use of social media and its linkage with the ability of students to think critically by deploying digital literacy skills such as searching, analysis, synthesis, evaluation, interpretation, creation, usage, communication, collaboration, among others (Kaeophanuek et al. 2018).

The layout of this book chapter consists of several sections. The first section, "literature review and development of hypotheses," defines each variable and discusses relevant published studies. Hypotheses were then developed based on evidence from the literature. In the next section, "methods," we thoroughly explain the participant selection process, the study procedures (including survey design), and the statistical data analysis techniques. In the next section, "discussion," result statements are presented with the support of existing relevant literature. And finally, "conclusion," will cover the research outcomes of this study that were attained through empirical results and the literature review.

Literature Review and Development of the Hypotheses

Use of Social Media and Critical Thinking

In higher education, critical thinking is defined as the student's ability to think, whether individually or in a group, and analyze, synthesize, and evaluate available information to make better decisions. In the digital era, the ability to think critically plays an important role in evaluation accumulation (Shcheglova et al. 2019). Bedenlier et al. (2020) found that Facebook-based activities increased the critical thinking abilities of students in social media learning environments when compared to face-to-face settings. Digital environments or spaces develop digital literacy skills and online engagement (Peters and Romero 2019) by having students search, analyze, and process information to then collaborate and share and finally disseminate, create, and innovate through digital tools. All these soft skills are directly linked to critical thinking (Kaeophanuek et al. 2018). Digital management skills help to engage and manage information, and improve theoretical knowledge and practical skills to overcome future challenges in a person's career (Janssen et al. 2019; Toplak et al. 2017). Social media platforms can enable students to develop their critical thinking by using digital literacy skills and vice versa in the digital era.

Use of Social Media and Engagement

In digital educational settings, learning platforms should be accessible and flexible such that all students can easily gain knowledge and improve their learning skills (Alghizzawi et al. 2019). In past years, social media has been transformed into a learning platform and is also used as an informal and formal channel of communication between individuals and groups of users. From an educational perspective, social media allows students to perform synchronous and asynchronous course activities. Facebook is an example of a social media platform that is used as a learning tool in the academic community (Knowles and Dixon 2016). The use of social media in online courses helps students to develop their soft skills through active collaborations and interactions (Gonzalez-Cacho and Abbas 2022). This collaboration and interaction can also be achieved through engagement. In an online classroom environment, students can engage with each other through social media-based activities. Educators are responsible for the design of online activities that contribute to the development of practical and theoretical skills of students. As a result, students will use creative activities for knowledge sharing and creation within a flexible learning space (Khoo 2019; Ilin 2021).

The Mediating Role of Engagement with the Use of Social Media and Critical Thinking

In the twentieth century, the concept of engagement among students started with Ralph Tyler and a new approach regarding the relationship between learning curricula and student learning (Axelson and Flick 2010; Manu et al. 2021). Since then, engagement is an important element in higher education and plays a central role in the learning processes (Ouyang and Dai 2022). Students develop/promote three types of engagement: behavioral, emotional, and cognitive. Different activities among students can promote behavioral engagement, whereas emotional engagement is based on reactions that appear during specifically designed activities, and finally, cognitive engagement clarifies complex ideas (Fredricks et al. 2004). Therefore, different types of engagement, similar to an individual's disposition, are made up of student emotions, cognition, and behavior (Kahu 2013) and it is important to understand this set of engagement elements. Engagement involves students in academic activities that result in high-quality outcomes in their studies (Mehdinezhad 2011). Therefore, students' attitudes of disposition, desire, and compulsion drive them to actively participate in learning activities (Mbodila et al. 2014) on social media. It is important for teachers to design innovative social media-based course activities to promote and enhanced the relationship between students' attitudes and academic achievements (Eastman et al. 2011). Social media provides an intellectual environment that can actively engage students and develop their critical thinking abilities. These learning activities allow teachers to monitor student engagement in activities related to academic commitments (Eom et al. 2016). Continuous engagement in students' intermediate social media activities relates to their personal interest in developing their soft skills, i.e., critical thinking (Sinatra et al. 2015; Uchidiuno et al. 2019).

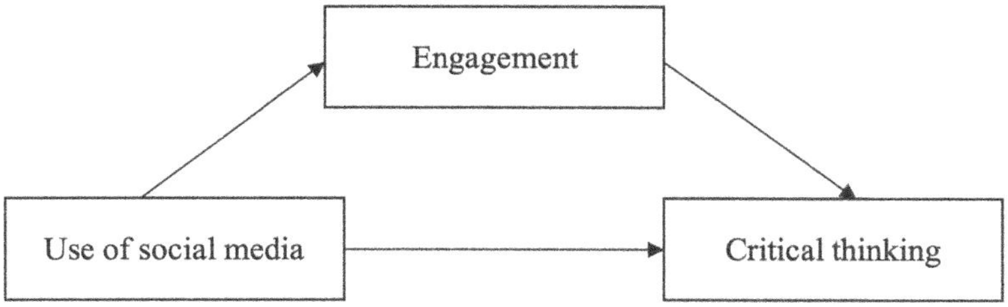

Fig. 7.1 Research model

Proposed Hypotheses

Based on the literature review, we proposed the following hypotheses:

H1: Use of social media is positively associated with critical thinking.
H2: Use of social media is positively associated with engagement.
H3: Engagement is positively associated with critical thinking.
H4: Engagement mediates the relationship between use of social media and critical thinking (Fig. 7.1).

Methods

Participants and Procedure

In this study, participants were undergraduate Architecture and Civil engineering students of Tecnologico de Monterrey, Puebla Campus of Mexico. This research is in line with the institutional COVID-19 policy of digital learning, known as HyFlex+Tec, which has the objective of helping students to continue their education through e-learning (Membrillo-Hernández et al. 2021). Tecnologico de Monterrey was the first Mexican University that suspended campus classes on March 17, 2020. From March 17 to 21, 2020 nearly 10,000 professors were trained in the use of digital platforms. Under the HyFlex+Tec on March 23, 2020, more than 90,000 students were able to continue their semester fully online. In Mexico, face-to-face classes were suspended at all educational levels since March 2020 with a gradual return to on-site classes in August 2021 (TEC 2022). Puebla campus is one of 26 campuses of Tecnologico de Monterrey, Mexico. In May 2021, via convenience sampling, we circulated a Google Form survey among students enrolled in the undergraduate Architecture and Civil engineering program. We used the university's official LMS platform, that is, Canvas (https://experiencia21.tec.mx) to distribute the survey. Survey respondents were already engaged in several social media-based course activities. During the COVID-19 pandemic, teachers designed a set of social media-based course activities to help students interact with their peers and share knowledge remotely. The objective of these social media-based academic activities is to promote critical thinking by deploying digital literacy skills to meet the future need for problem-solving and to improve decision-making.

The Online Survey

The online survey was based on four sections, (1) Demographic information of respondents i.e., age, gender, major, scholarship, and social media usage hours per day, (2) Use of social media, (3) Engagement, and (4) Critical thinking. The first section covers the confidentiality statement and includes questions related to age, gender, major, and the daily use of social media. The purpose of the confidentiality statement is to let respondents know the purpose of the research and also to inform them of their rights to privacy and security of their personal information. The second section contains four, five-point Likert-scale questions related to "use of social media" (Ali-Hassan et al. 2015). The third section includes four, five-point Likert scale questions related to "engagement" which were adopted from the work of Chan et al. (2020). The last section has six, five-point Likert-scale questions related to "critical thinking" which were adapted from research done by Gonzalez-Cacho and Abbas (2022). The questions of the survey were based on the different components of the RED Model of critical thinking developed by Pearson which is based on recognizing assumptions, evaluating arguments, and drawing conclusions (Chartrand et al. 2013).

Data Analysis

Data analysis was conducted using an open R-based statistical software, Jamovi. Jamovi software provides an advanced level of statistical testing options that can be used to analyze complex quantitative data. Before starting data analysis, we exported the Microsoft Excel dataset file to the Jamovi (.omv) file format. Then we assigned a label to each variable and a code to each registered response. The first step of data analysis was to apply a descriptive test to collect demographic information of each registered survey respondent. In the second step, we applied Pearson's r correlation test to verify the correlation between variables. Lastly, we applied a mediation test using the "medmod" module to test the mediating effects between variables. Therefore, our results are determined via statistical analyses, which will prove or disprove the proposed hypotheses.

Results

Descriptive Statistics

The following presents the study participant demographics of undergraduate Architecture and Civil engineering students. Fifty-four (80.6%) respondents were between the ages of 18–22 years, and the 13 remaining students were (19.4%) over the age of 22. Considering gender, 36 (53.7%) females and 31 (46.3%) males participated in the survey. Forty-nine (73.1%) students were majoring in Architecture and the remaining 18 (26.9%) students were majoring in Civil engineering. Forty-five (67.2%) of 67 students obtained a scholarship, and the remaining 22 (32.8%) were self-financed or without scholarship. Thirty-two students (47.8%) reported their social media usage as being 3–6 h per day, 30 (44.8%) students reported their usage as being less than 3 h a day, and the remaining five (7.5%) students reported their social media usage as being above 6 h a day.

Correlation Matrix

In Table 7.1, the correlation matrix shows that a relationship exists between the variables: use of social media, engagement, and critical thinking. Results of the Pearson's r correlation analysis suggest that "use of social media" has a positive correlation with "critical thinking" ($r = .687$, $p < .001$) and "engagement" is also positively correlated with "critical thinking" ($r = .732$, $p < .001$). Correlation values also suggest that the "engagement" variable is highly correlated with "critical thinking" when compared with "use of social media.".

Table 7.1 Correlation matrix

		Use of social media	Engagement	Critical thinking
Use of social media	Pearson's *r*	–		
	p-value	–		
Engagement	Pearson's *r*	0.762***	–	
	p-value	<.001	–	
Critical thinking	Pearson's *r*	0.687***	0.732***	–
	p-value	<.001	<.001	–

Note. Level of significance: ***$p < .001$

Testing the Mediation Effects

In mediation tests, we verified the direct, indirect, and total effects of variables. "Use of social media" is an independent variable, "critical thinking" is a dependent variable, while "engagement" is a mediator or mediating variable.

The first relationship test was between the independent variable, "use of social media, " and the dependent variable, "critical thinking" (see Table 7.2 and Fig. 7.2a). The results of the analysis confirm that the estimated value of the total effects of variables is significant (Estimate = 0.558, $p < .001$) and with a lower and upper limit of 95% confidence interval (C.I) is in the range between (0.4152–0.700), which confirms that relationship between the independent and dependent variables is significant. Therefore, the results support our first hypothesis, i.e., *H1: Use of social media is positively associated with critical thinking*.

In further analyses, we introduced the "engagement" variable as a mediator between "use of social media" and "critical thinking" (see Table 7.3 and Fig. 7.2b). In mediation tests, we found the first relationship between the independent variable "use of social media" and the mediation variable "engagement" is positively significant (Estimate = 0.787, $p < .001$), and second relationship between the mediation variable "engagement" and the dependent variable "critical thinking" is also positively significant (Estimate = 0.392, $p < .001$). Thus, the results from data analysis support our two hypotheses *H2: Use of social media is positively associated with engagement*, and also *H3: Engagement is positively associated with critical thinking*. As seen in Table 7.2, the value of indirect effects (Estimate = 0.308, $p < .001$) supports our hypothesis, *H4: Engagement mediates the relationship between the use of social media and critical thinking*. We can say that after introducing the mediator "engagement," we noticed that the value of direct effects changes to become ~~in~~significant. Because of this, our results show that partial mediation exists.

Table 7.2 Testing of mediating effects

				95% C.I. (a)				
Type	Effect	Estimate	SE	Lower	Upper	β	z	p
Total	Use of social media ⇒ Critical thinking	0.558	0.0727	0.4152	0.700	0.687	7.67	<.001
Indirect	Use of social media ⇒ Engagement ⇒ Critical thinking	0.308	0.0825	0.1466	0.470	0.380	3.74	<.001
Direct	Use of social media ⇒ Critical thinking	0.249	0.0998	0.0538	0.445	0.307	2.50	0.012

Note. Confidence intervals (C.I.) computed with method: Standard (Delta method)
Note. Betas are completely standardized effect sizes
Note. Level of significance: $^{*}p < .05$, $^{**}p < .01$, $^{***}p < .001$

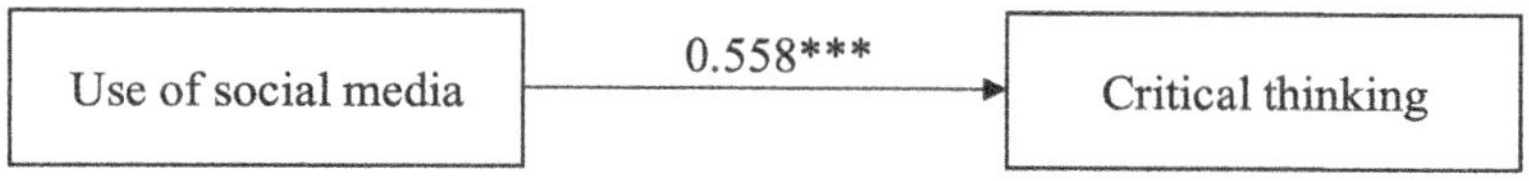

Fig. 7.2a Path estimate (total effects)

Table 7.3 Path estimate

Effect	Estimate	SE	95% C.I.		z	p
			Lower	Upper		
Use of social media → Engagement	0.787	0.0817	0.6268	0.957	9.63	<.001
Engagement → Critical thinking	0.392	0.0966	0.2023	0.581	4.05	<.001
Use of social media → Critical thinking	0.249	0.0998	0.0538	0.445	2.50	0.012

Note. C.I. Confidence Interval
Note. Level of significance: $*p < .05$, $**p < .01$, $***p < .001$

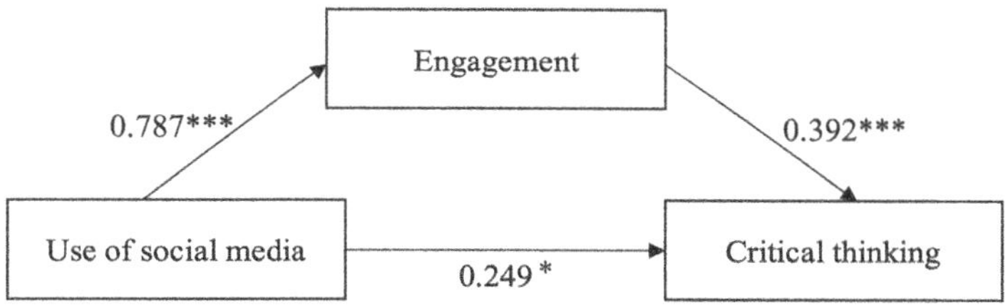

Fig. 7.2b Path estimate (indirect and direct effects)

Discussion

The purpose of this empirical study was to analyze the mediating role of engagement in the use of social media for the development of critical thinking abilities in undergraduate Architecture and Civil engineering students. To achieve this, we developed a theoretical research model and a list of proposed hypotheses. In this discussion section, we present, with the support of published literature, the results of this study.

The first finding of this study highlights the significant positive association that exists between the use of social media and students' critical thinking abilities as total effects. In the digital learning environment, students in higher education act as active learners whereby they use social media as a learning platform for active communication during the online activities of their courses. Students do this to exchange their ideas, share relevant skills, and knowledge with both peers and teachers (Pattanapichet and Wichadee 2015). Erdogdu (2022) mentioned in his research that effective use of an ICT-driven interactive platform provides a learning environment for the students to attain academic achievement along with improvements in their skills such as critical thinking.

The second set of findings suggests that the use of social media provides an important platform to engage students individually as well as in team-based course-related activities. In recent times, it has been noted that social media platforms, such as Facebook, YouTube, Twitter, and LinkedIn, provide an intellectual learning environment where anyone can express their ideas and knowledge on specific topics without any constraints (Shawky et al. 2019; Zhang et al. 2022). As such, a set of course-based activities designed by teachers help students to interact with each other during the COVID-19 pandemic, when most in-person educational activities came to a halt (Mershad and Said 2022; Haroon et al. 2020) Therefore, these online activities were helpful in assisting the students with improving their soft skills.

The third set of empirical findings also confirms that engagement is associated with the development of the critical thinking abilities of students. The framework of engagement is associated with academic research and the extracurricular engagement of students and their contributions to class discussions on concepts, issues, and solutions on specific course topics. These elements of the engagement framework are helpful beyond course structure and are also helpful for students in their extracurricular activities (Shcheglova et al. 2019). Online teaching methods are helpful for students wanting to acquire new skills and are desirous of achieving personal development goals as well as higher grades (Schmidt-Wilk and Lovelace 2017).

The final set of empirical findings affirms that engagement is a partial mediator between the use of social media (independent variable) and critical thinking (dependent variable). Anthony and Garner (2016) found that soft skills are attributed to students' learning abilities, which is important for them to improve communication, writing, and presentation skills through engagement in the digital learning environment. Effective use of technology provides flexibility in education and also encourages students to use internet-based applications for searching, collecting, and interpreting relevant information and knowledge for critical thinking (Pérez-Escolar and Canet 2022). Acceptance of and adopting emerging technologies in e-learning settings is beneficial for engagement which, in turn, positively influences student success and also has a long-term impact on their soft skills (Khlaif et al. 2022), which is necessary for successful career development.

Implication of the Study

This research study has valuable implications for policy and practice. Theoretically, this study contributes to the field by highlighting the role and effectiveness of social media in e-learning settings, particularly during the pandemic. As such, social media provides an intellectual space for students to (1) engage with their peers, (2) discuss and share relevant knowledge, and (3) develop skills for solving complex problems on specific topics within set timeframes. The practical implications of this empirical study can be used to guide educational institutions and their policies that seek to shift from in-person learning to online modalities where online course activities enable students to socially interact with each other. Online learning modalities also enable active engagement among students and their peers through the use of social media and also allow students to concentrate their focus on their studies and enhance their competencies for better academic performance and career development. Another implication is that this study highlights the teacher's role in designing online activities that promote digital literacy with the use of social media related to engagement and the development of soft skills.

Conclusion, Limitations, and Directions of Future Work

Considering the empirical evidence about the mediating role of engagement between use of social media and critical thinking abilities of undergraduate Architecture and Civil engineering students, we can conclude that the use of social media can play an important role in developing digital literacy and critical thinking. Studies show that the sole use of social media does not promote critical thinking but students' engagement through social media can develop critical thinking skills required to filter, analyze, synthetize, communicate, and collaborate effectively using social media platforms. In a time when face-to-face education is transitioning to a hybrid model of instruction, social media is a powerful tool that could be used in formal and informal learning contexts. During the pandemic, many institutions transformed their education systems to online-based learning. As such, most of the degree programs designed their course content based on online activities such as lectures, presentations, and online individual or team-based project activities. The use of social media as a learning platform helps undergraduate students actively participate and engage in course activities by engaging with their peers and teachers. Online learning platforms also assist in the learning process as students share relevant information and knowledge on specific topics. This set of online activities engages students to develop their critical thinking abilities for better decision-making and future career development.

Along with the significant contributions of this study, we also noted limitations. The limitation of this research is that it was conducted among undergraduate Architecture and Civil engineering students on one campus of a university. Possibilities for future research can include multidisciplinary study with a different department of the university or with engineering students from a different campus as it will be helpful to get better insights regarding the role of social media platforms for the development of theoretical and practical skills of students.

References

Ahmad AR, Jameel AS, Raewf M (2021) Impact of social networking and technology on knowledge sharing among undergraduate students. Int Bus Educ J 14(1):1–16. https://doi.org/10.37134/ibej.vol14.1.1.2021

Ahmadi DMR (2018) The use of technology in English language learning: a literature review. Int J Res English Educ 3(2):115–125. https://doi.org/10.29252/ijree.3.2.115

Alghizzawi M, Habes M, Salloum SA, Ghani M, Mhamdi C, Shaalan K (2019) The effect of social media usage on students' e-learning acceptance in higher education: a case study from The United Arab Emirates. Int J Inf Technol Lang Stud 3(3):13–26

Ali-Hassan H, Nevo D, Wade M (2015) Linking dimensions of social media use to job performance: the role of social capital. J Strateg Inf Syst 24(2):65–89. https://doi.org/10.1016/j.jsis.2015.03.001

Al-Rahmi WM, Othman MS, Yusuf LM (2015) Effect of engagement and collaborative learning on satisfaction through the use of social media on Malaysian higher education. Res J Appl Sci Eng Technol 9(12):1132–1142. https://doi.org/10.19026/rjaset.9.2608

Anthony S, Garner B (2016) Teaching soft skills to business students: an analysis of multiple pedagogical methods. Bus Prof Commun Q 79(3):360–370. https://doi.org/10.1177/2329490616642247

Armstrong A, Georgas H (2006) Using interactive technology to teach information literacy concepts to undergraduate students. Ref Serv Rev 34(4):491–497. https://doi.org/10.1108/00907320610716396

Axelson RD, Flick A (2010) Defining student engagement. Change Mag High Learn 43(1):38–43. https://doi.org/10.1080/00091383.2011.533096

Balasooriya I, Mor E, Rodríguez ME (2018) Understanding user engagement in digital education. In: Learning and collaboration technologies. Learning and teaching. Springer, Cham, pp 3–15. https://doi.org/10.1007/978-3-319-91152-6_1

Bedenlier S, Bond M, Buntins K, Zawacki-Richter O, Kerres M (2020) Facilitating student engagement through educational technology in higher education: a systematic review in the field of arts and humanities. Australas J Educ Technol 36(4):126–150. https://doi.org/10.14742/ajet.5477

Bhardwaj P, Gupta P, Panwar H, Siddiqui MK, Morales-Menendez R, Bhaik A (2021) Application of deep learning on student engagement in e-learning environments. Comput Electr Eng 93:107277. https://doi.org/10.1016/j.compeleceng.2021.107277

Chan I, Lau Y-y, Lee WSW (2020) Adoption of knowledge creation model in team-based project to support student engagement. Int J Innov Creativity Chang 14(11):1–15

Chartrand J, Ishikawa J, Flander S (2013) Critical thinking means business: learn to apply and develop the new #1 workplace skill. TalentLens. https://us.talentlens.com/content/dam/school/global/TalentLens/us/whitepapers/critical-thinking-means-business.pdf

Clark M, Fine MB, Scheuer C-L (2017) Relationship quality in higher education marketing: the role of social media engagement. J Mark High Educ 27(1):40–58. https://doi.org/10.1080/08841241.2016.1269036

Dabbagh N, Kitsantas A (2012) Personal learning environments, social media, and self-regulated learning: a natural formula for connecting formal and informal learning. Internet High Educ 15(1):3–8. https://doi.org/10.1016/j.iheduc.2011.06.002

Eastman JK, Iyer R, Eastman KL (2011) Business students' perceptions, attitudes, and satisfaction with interactive technology: an exploratory study. J Educ Bus 86(1):36–43. https://doi.org/10.1080/08832321003774756

Eom SB, Wen HJ, Ashill N (2016) The determinants of students' perceived learning outcomes and satisfaction in university online education: an update*. Decis Sci J Innov Educ 14(2):185–215. https://doi.org/10.1111/dsji.12097

Erdogdu F (2022) ICT, learning environment and student characteristics as potential cross-country predictors of academic achievement. Educ Inf Technol. https://doi.org/10.1007/s10639-021-10848-x

Fredricks JA, Blumenfeld PC, Paris AH (2004) School engagement: potential of the concept, state of the evidence. Rev Educ Res 74(1):59–109. https://doi.org/10.3102/00346543074001059

Gonzalez-Cacho T, Abbas A (2022) Impact of interactivity and active collaborative learning on students' critical thinking in higher education. IEEE Rev Iberoam Tecnol del Aprendiz 17(3):254–261. https://doi.org/10.1109/rita.2022.3191286

Haq IU, Anwar A, Rehman IU, Asif W, Sobnath D, Sherazi HHR, Nasralla MM (2021) Dynamic group formation with intelligent tutor collaborative learning: a novel approach

for next generation collaboration. IEEE Access 9:143406–143422. https://doi.org/10.1109/access.2021.3120557
Haroon Z, Azad AA, Sharif M, Aslam A, Arshad K, Rafiq S (2020) COVID-19 era: challenges and solutions in dental education. J Coll Physicians Surg Pak 30(10):S129–S131. https://doi.org/10.29271/jcpsp.2020.Supp1.S129
Ilin V (2021) The role of user preferences in engagement with online learning. E-Learn Digit Media:1–20. https://doi.org/10.1177/20427530211035514
Janssen EM, Mainhard T, Buisman RS, Verkoeijen PP, Heijltjes AE, van Peppen LM, van Gog T (2019) Training higher education teachers' critical thinking and attitudes towards teaching it. Contemp Educ Psychol 58:310–322. https://doi.org/10.1016/j.cedpsych.2019.03.007
Kaeophanuek S, Jaitip N-S, Nilsook P (2018) How to enhance digital literacy skills among information sciences students. Int J Inf Educ Technol 8(4):292–297. https://doi.org/10.18178/ijiet.2018.8.4.1050
Kahu ER (2013) Framing student engagement in higher education. Stud High Educ 38(5):758–773. https://doi.org/10.1080/03075079.2011.598505
Khlaif ZN, Sanmugam M, Ayyoub A (2022) Impact of technostress on continuance intentions to use mobile technology. Asia Pac Educ Res. https://doi.org/10.1007/s40299-021-00638-x
Khoo BK (2019) Mobile applications in higher education: implications for teaching and learning. Int J Inf Commun Technol Educ 15(1):83–96. https://doi.org/10.4018/ijicte.2019010106
Knowles M, Dixon D (2016) Guidance techniques to improve student engagement in critical reflection regarding the preparation of technical reports. Stud Engage Exp J 5(1). https://doi.org/10.7190/seej.v5.i1.104
Le B, Lawire GA, Wang JTH (2022) Student self-perception on digital literacy in STEM blended learning environments. J Sci Educ Technol. https://doi.org/10.1007/s10956-022-09956-1
Mafarja N, Zulnaidi H, Fadzil HM (2022) Using reciprocal teaching strategy to improve physics students' critical thinking ability. Eurasia J Math Sci Technol Educ 18(1):em2069. https://doi.org/10.29333/ejmste/11506
Malik MJ, Ahmad M, Kamran MR, Aliza K, Elahi MZ (2020) Student use of social media, academic performance, and creativity: the mediating role of intrinsic motivation. Interact Technol Smart Educ 17(4):403–415. https://doi.org/10.1108/ITSE-01-2020-0005
Manu BD, Ying F, Oduro D, Boateng SA (2021) Student engagement and social media in tertiary education: the perception and experience from the Ghanaian public university. Soc Sci Humanit Open 3(1):100100. https://doi.org/10.1016/j.ssaho.2020.100100
Mbodila M, Ndebele C, Muhandji K (2014) The effect of social media on student's engagement and collaboration in higher education: a case study of the use of Facebook at a South African university. J Commun 5(2):115–125. https://doi.org/10.1080/0976691x.2014.11884831
Mehdinezhad V (2011) First year students' engagement at the university. Int Online J Educ Sci 3(1):47–66
Membrillo-Hernández J, García-García R, Lara-Prieto V (2021) From the classroom to home: experiences on the sudden transformation of face-to-face bioengineering courses to a flexible digital model due to the 2020 health contingency. In: Educating engineers for future industrial revolutions. Springer, Cham, pp 488–494. https://doi.org/10.1007/978-3-030-68201-9_48
Mershad K, Said B (2022) DIAMOND: a tool for monitoring the participation of students in online lectures. Educ Inf Technol. https://doi.org/10.1007/s10639-021-10801-y
Moon SJ, Bai SY (2020) Components of digital literacy as predictors of youth civic engagement and the role of social media news attention: the case of Korea. J Child Media 14(4):458–474. https://doi.org/10.1080/17482798.2020.1728700
Ouyang F, Dai X (2022) Using a three-layered social-cognitive network analysis framework for understanding online collaborative discussions. Australas J Educ Technol 38(1):164–181. https://doi.org/10.14742/ajet.7166
Pattanapichet F, Wichadee S (2015) Using space in social media to promote undergraduate students' critical thinking skills. Turk Online J Dist Educ 16(4):38–49. https://doi.org/10.17718/tojde.94170

Pérez-Escolar M, Canet F (2022) Research on vulnerable people and digital inclusion: toward a consolidated taxonomical framework. Univ Access Inf Soc. https://doi.org/10.1007/s10209-022-00867-x

Peters M, Romero M (2019) Lifelong learning ecologies in online higher education: students' engagement in the continuum between formal and informal learning. Br J Educ Technol 50(4):1729–1743. https://doi.org/10.1111/bjet.12803

Radovanović D, Hogan B, Lalić D (2015) Overcoming digital divides in higher education: digital literacy beyond Facebook. New Media Soc 17(10):1733–1749. https://doi.org/10.1177/1461444815588323

Radovanović D, Houngbonon GV, Le Quentrec E, Isabwe GMN, Noll J (2020) Digital infrastructure enabling platforms for health information and education in the global south. In: Digital inequalities in the global south. Springer, pp 199–222. https://doi.org/10.1007/978-3-030-32706-4_10

Schmidt-Wilk J, Lovelace K (2017) Helping students succeed through engagement and soft skills development. Manag Teach Rev 2(1):4–6. https://doi.org/10.1177/2379298116687684

Shawky S, Kubacki K, Dietrich T, Weaven S (2019) Using social media to create engagement: a social marketing review. J Soc Mark 9(2):204–224. https://doi.org/10.1108/JSOCM-05-2018-0046

Shcheglova I, Koreshnikova Y, Parshina O (2019) The role of engagement in the development of critical thinking in undergraduates. Voprosy Obrazovaniya/Educ Stud Moscow 1:264–289

Sherman K (2013) How social media changes our thinking and learning. Lang Teach 37(4):8. https://doi.org/10.37546/jalttlt37.4-3

Sinatra GM, Heddy BC, Lombardi D (2015) The challenges of defining and measuring student engagement in science. Educ Psychol 50(1):1–13. https://doi.org/10.1080/00461520.2014.1002924

TEC (2022) Classes and experience for 90,000 students: challenge faced by Tec. https://conecta.tec.mx/en/news/national/education/classes-and-experiences-90000-students-challenge-faced-tec

Toplak ME, West RF, Stanovich KE (2017) Real-world correlates of performance on heuristics and biases tasks in a community sample. J Behav Decis Mak 30(2):541–554. https://doi.org/10.1002/bdm.1973

Uchidiuno J, Yarzebinski E, Keebler E, Koedinger K, Ogan A (2019) Learning from African classroom pedagogy to increase student engagement in education technologies. In: COMPASS '19: proceedings of the 2nd ACM SIGCAS conference on computing and sustainable societies, Accra, Ghana, 2019. Association for Computing Machinery, New York, pp 99–110. https://doi.org/10.1145/3314344.3332501

Zhang Z, Ning M, Wu Z, Liu C, Li J (2022) Construction of a competition practice system for robotics engineering based on a competency model. Comput Appl Eng Educ. https://doi.org/10.1002/cae.22491

Chapter 8
Tales of Visibility in TikTok: The Algorithmic Imaginary and Digital Skills in Young Users

Elisabetta Zurovac, Giovanni Boccia Artieri, and Valeria Donato

Introduction

Our lives are increasingly affected by algorithms that select information and regulate processes of interaction, providing us with a set of choices. As intermediaries in the distribution of information, they intervene in decisional processes and regulate access to several fundamental social spheres including work, education, or justice (O'Neil 2016) by providing users with relevant information that is required to participate in public life (Gillespie 2014). However, these mechanisms remain opaque to users. While they can encourage social connections (van Dijck 2013) and help us to orient ourselves in the complex enormity of information, they can also produce harmful or discriminatory results. For example, it has been shown that search algorithms can present biases that marginalize minorities, women and people with lower income (Noble 2018), thus contributing to the production and reproduction of stereotypes and discrimination. As algorithms play such a fundamental role on social media, where adolescents' digital skills are forged, there is consequently an urgency to investigate their experience when encountering algorithms.

E. Zurovac (✉) · G. Boccia Artieri · V. Donato
Department of Communication, Humanities and International Studies (DISCUI), University of Urbino Carlo Bo, Urbino, Italy
e-mail: elisabetta.zurovac@uniurb.it

D. Radovanović (ed.), *Digital Literacy and Inclusion*,
https://doi.org/10.1007/978-3-031-30808-6_8

Algorithmic Awareness and Practices on TikTok

Among the digital platforms that are most used by adolescents, contributing thus to their digital skills, TikTok is currently one of the most popular (Statista 2021). Moreover, its "engineered playfulness and performativity" (Savic and Albury 2019) is sustained by a highly intuitive set of video editing tools. Recent analyses carried out on TikTok (Bhandari and Bimo 2020; Francisco and Ruhela 2021; Krutrök 2021) show how this platform produces a model of sociability that is different from other social media-generated ones. TikTok is characterized by online interaction through an "algorithmized self": a type of public presentation oriented by a highly personalized algorithm (Francisco and Ruhela 2021) that is heavily informed by and directed towards the individual instead of an "audience" and thus guided by intrapersonal engagement rather than interpersonal engagement.

As Krutrök (2021) has argued, users develop socio-technological skills not through previously acquired digital skills related to algorithms, but from progressive exposure to the platform and to the practices they observe within it. In other words, users become aware of the existence of an algorithm, and by repeatedly observing and performing practices within that platform, they adapt themselves accordingly (Zulli and Zulli 2020). Observing repeated practices is so-called algorithmic gossiping (Bishop 2019), representing the way users spread and understand the logic of algorithms.

As noted by Karizat et al. (2021), the algorithmic awareness seems to be crucial to the user's experience within the TikTok: it allows the development of algorithmic folk theories (Eslami, et al. 2016) that lets users behave, engage and also resist the platform's structure. Folk theories are relevant within social media users due "*to the highly personalised, continuously changing, and relatively obscure nature of the algorithmically-driven systems for content curation and moderation*" (Chung 2020, p. 11).

However, a full understanding of the TikTok algorithm still does not allow users to avoid problematic contents (Harrigera et al. 2022) or escape from an already determined information literacy (Radovanović et al. 2015; Haider and Sundin 2021).

Taking into account this relevant literature, we focus our analysis on the algorithmic practices rather than the algorithmic knowledge as an appropriate and exhaustive understanding of how an algorithm works (in general) is not possible to obtain. What a user can do instead is imagine, learn by observing other users and try out tricks and/or practices demonstrating algorithmic awareness.

Building on the analysis of algorithmic practices (Boccia Artieri 2020) enacted on TikTok, this chapter aims to explore the relationship between the digital skills of adolescents and their algorithmic imaginary (Bucher 2017) as it is tied to the platform. Bearing this in mind, as will be illustrated in the following section, these practices depend on which "side of TikTok" a user is situated (Krutrök 2021). As the platform's algorithm pushes users into specific communities and niches, as Gillespie has indicated, the algorithm "shapes a public's sense of itself" (2014, p. 3) that represents the basis of their digital self-representation.

It is for this reason that qualitatively studying such performances of the self, embedded within a platform that is characterized by a sophisticated system of algorithmic recommendations (Vázquez-Herrero et al. 2020), also represents a way to understand how digital skills are employed in relation to the algorithmic imaginary.

A Carousel of Dreams and Nightmares: TikTok's FYP, Between Content Suggestion and Moderation

While other social media generate users' feeds through their social connections, TikTok can determine user preferences by gathering other personal information, such as location, gender and the device being used (Francisco and Ruhela 2021). As if that were not enough, TikTok is also able to identify small similarities between the most popular videos in order to construct a user profile to whom the platform will suggest content accordingly (Yu 2019; Smith 2021). So, while on other platforms the user feed is populated by content shared by a pre-existing network of followed accounts, TikTok's main page—the "for you" page (FYP)—is determined by its recommendation system. As a result, users are shown content based on previous preferences together with a carousel of possible choices for assessing future content on display (Klug et al. 2021). This carousel, according to TikTok (2020), is built on the main features of the app: it picks up content that uses the most popular sounds, effects and hashtags. The latter dimension needs to be addressed further. Though not as central as hashtags on other platforms from a user perspective (Yu 2019), textual information on TikTok is very valuable to the recommendation system as it uses hashtags and captions in order to evaluate content. The "LIVE page" of TikTok, which collects live streamed videos, is connected to the FYP as follows: while scrolling, users may encounter suggested LIVEs to interact with, and if a user decides to "enter" into it, TikTok opens a dedicated scrolling page. The LIVE shown on the FYP functions as a hook that is selected specifically for the user, while the other, specific "LIVE page" shows potentially all of the LIVE feeds being streamed at that moment. Creators that gain access to the LIVE function are those that have reached 1000 followers and are older than 16.

Although it appears particularly accurate in its recommendation and therefore its evaluation system (Smith 2021), the platform also raises several concerns. The lack of pre-existing social bonds and the use of a pseudonym allow users to express themselves more freely, within a space of semi-anonymity where the context collapse (Marwick and Boyd 2011) is often avoided. As a consequence, TikTok users are exposed to strangers and encouraged by the recommendation system towards their specific niche or subculture (Wall Street Journal 2021).

As Weimann and Masri (2020) note, the platform seems unable to enforce its guidelines and, most importantly, exposure to one problematic video potentially leads the algorithm logic to show further like content. This awareness should be at the basis of the defense mechanism used to "escape" from the bubbles in which users are placed.

Research Questions and Methodology

While the affordances of the platform seem to facilitate the diffusion and visibility of content, users still need to develop in-depth knowledge of TikTok's logic to benefit from it in terms of success and visibility. Users must similarly develop specific skills if they want to elude the platform's scrutiny—in terms of both content moderation and suggestions. Also, TikTok involves the users within one niche or another, framing the experience that occurs within the platform. In fact, it brings forth a highly localized and shared sociality structured in sub-communities that may not be accessible (nor visible) to non-members (Abidin 2021a). These dynamics are particularly relevant when thinking about young people, as they represent the vast majority of TikTok's audience and the most vulnerable public in terms of passively absorbing knowledge.

This leads to the research questions that this chapter aims to answer:

(RQ1) In which ways do young users employ or trick the algorithm on TikTok?
(RQ2) Do users employ digital skills while producing content on TikTok; if so, how?

In order to answer these questions, we adopt a qualitative methodological approach. Specifically, we consider digital ethnography (Hine 2015) to be the most efficient method to gain a longitudinal perspective aimed at understanding the complexity of the platform and the content (Schellewald 2021). The observation period lasted from July 2020 to November 2021, and each of the three researchers, located in different parts of Italy, used its personal account on a daily basis, for the extent of at least an hour, employing a spreadsheet as a diary and for data storage.

During this period, each researcher observed different practices and explored different sides of TikTok, participating in the platform as a user would do. In doing so, each researcher was able to develop a deep contextual understanding of the content. As noted by Schellewald (2021), the typologies of content users may encounter at the beginning of their experience are: comedic, documentary, communal (i.e., memes, trends), explanatory, interactive and meta. Due to the platforms' complexity and algorithmic sensitivity to one's preferences each researcher obtained different results within her/his FYP.

This approach led to the identification, as the so-called "pop-up moments" (Hine 2015), of three case studies, related to three specific elements occurring within TikTok: sound, hashtags and LIVEs. These cases were evaluated qualitatively, so as to answer RQ1. Specifically, understanding TikTok as a "meme breeding ground" (Martin 2019) while analysing content, we took into consideration the memetic dimensions suggested by Shifman (2013): content, form and stance. Each dimension was adjusted for the field and grounded in the knowledge developed within TikTok. To address RQ2, we draw from and build on the systematization of studies of digital skill domains presented by Haddon et al. (2020). Specifically, we evaluated, in each case, the presence of:

- Social interaction skills, defined as the capacity to understand the conventions of social communication

- Content creation skills, defined as the capacity to create suitable content and editing skills
- Ethical behaviour online, defined as the capacity to carefully explore content and/or engage with others

The data collected qualitatively refer to public videos produced by users declaring their age (13–19) within their self-description displayed on their profile.

"Sorry I Forgot TW!": Managing Boundaries and Self-Expression in #Traumatok

In March 2021, one of the researchers was consistently exposed to videos that narrate users' personal traumas, noting that almost every video presented the hashtag #Traumatok. Considering the hashtag as an instrument that can maximize visibility and define the field, the researcher moved to explore it more closely. The videos, whose duration varied from 15 s to 3 min, have been collected and analysed between March and November 2021.

This side of TikTok is populated by users who are united by the desire or the necessity to share traumatic experiences concerning abuse, childhood trauma, addiction, mental health and/or gender-related trauma.

Given the specificity of these topics and the age of the users involved, the main concerns revolve around (a) the risk of exposure to potentially harmful content; and (b) the risk of oversharing personal information.

As Weimann and Masri (2020) have demonstrated, TikTok does not always succeed in the complete removal of problematic content. In fact, the platform employs an accurate data mining system to apply the shadow ban, but users can dodge this mechanism by changing the spelling of triggering words. Using combinations of letters, symbols, emojis and numbers (e.g., "mol3$t3d") is a common practice among the young population of Traumatok, when taking into consideration any of the textual parts of the content. Another such "precautions" related practice was identified in this context, albeit one which, differently, did not involve solely the algorithm.

Well aware of the potential visibility of content that could appear randomly in the FYP of unaware and unprepared individuals, users sometimes apply a warning signal in the form of letters "TW," that is, "trigger warning." In this way, responsibility for problematic content is delegated to the recommendation system (as it promotes videos for the FYP) and to the audience (immediately made aware by this warning). This tactic is an example of positive online behaviour intertwined with an awareness of unexpected users, and it demonstrates the development of a social norm. This codified disclaimer represented a marginal practice, however, despite the topics presented in the content (such as sexual assault or suicide). In one case, a creator writes in the comments "*SORRY I FORGOT A TW!! TW: joke about committing non-alive*"; this shows a desire to respect others' boundaries but also a carefree attitude when this norm is violated.

Apropos of boundaries, the other principal concern that can arise while exploring Traumatok is the management of personal information when narrating a traumatic experience. It should be noted that this context is intended as a "safe space" for the collective sharing of private traumas. Users feel free to express themselves: on the one hand, there is a community of individuals providing mutual support in the comments; on the other, since these communities are populated by strangers using pseudonyms, the tension caused by the "context collapse" experienced on other social media is relieved. That said, crossing the line of what should not be shared publicly is tricky to assess. Indeed, some information, if displayed, could potentially represent a privacy risk. Our analysis nevertheless reveals that these details are rarely given, and therefore that the young creators of Traumatok actively manage the boundaries of what is safe to share.

With regard to the video (self-)portrayal of the users, the vast majority adopts a curated image: even when talking directly about trauma, appearance has much relevance. This not only shows an awareness of "being in public" but also an acquired competence about the algorithm. In fact, as Melonio and colleagues (2020) have illustrated, for instance, TikTok has had issues with the "beauty stigma," as it seems to provide greater visibility to good-looking users. Indeed, appearance is so relevant on the platform that sometimes even the commenters point out the physical features of the creators, as well as mention the heart-wrenching topics of the videos.

The playful way of being on TikTok (Savic and Albury 2019) appears to dominate, encouraged in particular by the platform's memetic design (Zulli and Zulli 2020); this characteristic appears to affect even the more controversial niche of Traumatok. Our analysis reveals that most videos utilize lip-synching, popular sound effects and dark humour as coping mechanisms. For example, this is the case of a user that shows her face in blurred make-up after crying and uses as background sound some disturbing vocal messages that her ex sent her. While the audio goes on and the transcription of the messages appears on the screen, she mimics a dance following the words, as has been seen in any other performative TikTok content. Another user, for instance, employs audio called "step on the gas," which comes from a Playstation popular video game (Barradale 2021) but it is often used within #traumatok to present evidence of different kinds of abuse. In this specific example, while lip-syncing the word she writes on the screen about the fact that her mother forcing her into dieting made her develop a severe eating disorder. In this sense, "step on the gas" highlights in a "funny" way the tragic exaltation of her condition. In most cases, this creates an ad hoc performativity, but a minority of users also take part in more mainstream trends, reshaping the original meaning to make it coherent with the narrative costumes of this niche area. The adoption of sound and the aforementioned playful creativity, while talking about trauma, are elements that lead us to define this content as "commodified." That means videos are created to be relatable: they include no personal information, a generic description of a situation and correct use of the platform's editing tools. The success of this kind of content is attested by the comments, which typically just mention other users, recalling their attention towards something that is worthy of being seen.

Nevertheless, we should not over-simplify Traumatok as a demonstration of the desire to gain visibility through the spectacularization of pain. Rather, the platform intertwines the dimension of "popularity" with the desire to be seen and heard, and therefore the need to reclaim one's power and one's voice in the narration of trauma and abuse.

Given this evidence, we may assert that users understood how to create effective content (content creation skills), also with regard to their appearance (social interaction skills) therefore taking into account the algorithmic imagery they have developed (in regard to that we refer specifically to the modified spelling of sensitive words). Moreover, when thinking about the use of "trigger warning," we can also deduce that some of the users present online ethical behaviour skills, as they prevent other users from being exposed to potentially disturbing content. At the same time, thinking about the performative aspects of the contents related to such themes may raise questions about self-exposure and spectacularization of trauma that may be problematic in terms of ethics.

Sounds Attractive: The Dynamics of Visibility in the Breonna Taylor Trend

In May 2020, the protests promoted by the Black Lives Matter movement pushed the spotlight onto another case of police brutality that happened in March 2020, concerning the murder of 26-year-old Breonna Taylor. Beyond mainstream media coverage, protesters used social media to sensitize, engage and cooperate through the hashtags #blacklivesmatter and #BLM, and the cause reached huge visibility and worldwide participation.

In July 2020 one of the researchers noted the repetition of numerous similar videos using the same song, titled "I Need You To" by the musician Tobe Nwigwe. The 44-second song, released on Instagram on 6 July 2020, spread through TikTok, confirming that it represents a new space for young users to distribute political content (Serrano et al. 2020). In the video accompanying the song, the call to action for the viewers is represented by the title and reinforced by the transcription of the lyrics "arrest the killers of Breonna Taylor."

Due to the prominence of this song, the researcher used the possibility given by the platform to search for content following this sound, from July to September 2020.

The majority of the analysed videos presented the same scheme: a kind of engaging or distracting introduction interrupted by a close-up of the user lip-synching to the verse "arrest the killers of Breonna Taylor," while this text appeared on the screen. The main techniques used to make the introduction include click-bate topics, which in turn are reaffirmed in the captions. Hashtags are not always attached to the video; if present, in the majority of cases, they do not relate to Breonna Taylor. In order to tease the audience, users mostly pretend to be sharing (1) gossip, (2) personal secrets or (3) life hacks.

While pleasing the curiosity of the audience serves as a hook for the watcher to stick around for enough time to gain visibility Smith 2021, the absence of revealing hashtags is a way to avoid being spotted and shadow-banned by the platform. This absence of coherent textual indicators (hashtags and captions) invokes two different actors: the audience and the algorithm. On the one hand, it does not allow the audience to grasp the ulterior motive of the video and on the other, it prevents the algorithm from automatically pushing aside this content. In a few cases, the fear of being erased or intentionally blocked from emerging on the FYP is clearly expressed within text juxtaposed on the video or claimed in the caption (e.g., "watch it before TikTok takes it down"). These are extremely rare cases, though they demonstrate nonetheless that some users have at least a partial recognition of the depth of the platform's control. Indeed, after initially removing from the FYP videos dealing with protests (Harris 2020), TikTok (2020) backtracked, declaring that this was a mistake and that it would be corrected. With this in mind, it suggests that young users of the platform were not completely aware of this information, even if it was circulating publicly.

Exploring the phenomenon more deeply, the objective of gaining or providing visibility can be observed through the tactics used by the creators and by the audience.

In order to produce an effective video, users exhibit two main strategies suggested both by the affordances of the platform and by the algorithm: (1) a curated self-presentation, and (2) the correct use of filters. With the notion of a curated self-presentation, we refer to the overall appearance (i.e., the presence of make-up, a refined outfit and hairstyle) and the way in which an individual acts in front of the camera (e.g., self-confident, flirty). From this perspective, filters can of course be considered as a tool to enhance one's appearance, but, relevantly, they also represent an element that the algorithm evaluates positively. Here we can distinguish two predominant kinds of filters: those improving users' looks (white teeth, smooth skin) and those refining the content through visual effects (such as a green screen or a rain filter).

To consider the tactics employed by the audience, besides likes and sharing, we focused on the comment section. In fact, the knowledge that commenting would result in a better performance for the content, users explicitly wrote messages such as "commenting for the algorithm" or just "algorithm." This shows their interest in sustaining the message, by increasing the visibility of the video and exploiting their grasp of TikTok strategies.

Analysing the comments also led us to reflect on content reception. Being a TikTok trend evidently indicates growth in popularity, but it also means greater exposure to criticism. The Breonna Taylor trend, while giving voice to a relevant social and political movement, also represents a way to obtain views. This short circuit is at the basis of the debate around networked activism (Boccia Artieri 2021), which has gained, over the last decade, an infamous meaning. In fact, the label "performative activism" (Alperstein 2021) has been used by the audience as an accusation of content creators, implying that participating in the trend represented no more than a bandwagon and an opportunity for personal success. Of course, the use of filters and a curated self-presentation can be seen as reinforcing this view, but the main evidence of superficial participation was (for the public) the absence of a concrete call to action, such as promotion of the online petitions promoted by the BLM movement.

In conclusion, we may assert that users understood how to create effective content (content creation skills) in terms of the attention economy and therefore visibility. Also, the use of effective baits during the trend with regard to their imagined audience demonstrates the employment of social interaction skills. This imagined audience and the trend itself also show that they are taking into account the algorithmic imagery they have developed. However, when thinking about the "memefication" of a tragedy such as that which occurred to Breonna Taylor and her family, it is legit to question whether or not this kind of participation is to be considered ethical behaviour.

The Silence of Ordinary LIVEs: Social Interaction Skills Between Reward Practices and Expected Behaviours

While most of the research on TikTok focuses on sounds (Wright 2021), hashtags and communities (Krutrök 2021), the aspect of the Live page has received little attention. At the end of July 2021, TikTok (2021) improved its live streaming service, launching, among other new features, the inclusion of live content within the FYP, enabling easier access to LIVEs and therefore wider visibility. Between August and September 2021, the researchers encountered LIVE streaming suggestions multiple times, so the explorative research of this element began in September 2021 and lasted until December of the same year.

The main affordances of the TikTok LIVEs are: the possibility to use filters; the chance to receive "gifts" from the audience (expressed in stickers with different values of TikTok coins); gaining comments; and choosing followers as moderators. Moreover, TikTok LIVEs are ephemeral (they cannot be saved and posted) and they are addressed to a wide public (the audience can be potentially anyone). Contrarily to the products observed in the two previous cases, LIVEs are, by definition, immediate. As a result, this content is streamed as it is: without complex editing, a plot, or other heavily, visible manipulation of the take. This produces a sort of "raw footage" that is dissimilar to the usually curated content of TikTok. Indeed, it rapidly became clear in our results that the main characteristic of streaming is its "ordinary" dimension: they often depict the household, if not other routine locations (e.g., school, the bus stop, the park) where life occurs without no out of the ordinary events. In this way, LIVEs reinforce the idea of immediacy: they offer an invitation to share and engage with such irrelevant moments. Creators are well aware of their audiences, and they expect from them an active response. Indeed, microcelebrity practices (Zurovac and Boccia Artieri 2019) are interiorized by these users: they expect to be asked questions and provided with topics of conversation by the audience. However, of course, merely "being there" does not necessarily translate into "being interesting," and at times LIVEs are dominated by silence.

Although ordinary LIVEs are prevalent, there are some examples of more refined content that we label as "performative," albeit faced with some difficulty to specify on this definition even in view of certain distinguishing formats. As in the case of the "get ready with me" LIVEs, where creators—mimicking a format from YouTube (Hill 2019)—fill up an insignificant moment of their time while preparing for some other activity. We also find codified formats that are TikTok native, such as the "studying LIVE." This is visible in the case of a 19-year-old girl with 190 thousand followers, who films herself at the desk in her bedroom while doing homework and listening to music. She explained the context and the engagement rules (i.e., "live studying session 1:30 hour and 5 minutes pause for chatting") on a poster positioned next to her. Moreover, in order to capitalize visibility, on the poster she includes (1) her Instagram handle, inviting "adds"; (2) an invitation to interact with the LIVE by sending hearts; and (3) a subtle request to send TikTok gifts, to support her.

Though the platform's guidelines state that users are not allowed to ask for gifts or donations of any kind, this last request appears to be common practice. Broadly, such casual invitations can also evolve into "forced interaction" when creators openly show their rules of engagement, where any interaction (such as mentioning or adding someone from the audience) has its value in TikTok gifts. Putting effort into this capitalized attention is further demonstrated by the fact that curated self-presentation appears predominant. In other words, even when the creator is recording a studying session, for instance, often their make-up, outfit and hair are nonetheless well put-together. The researchers also noted that the attention to appearance is again often vocalized by the audience, as well: no matter the content of the LIVE (even in the case of a heartbroken 15-year-old boy crying) the appearance of the TikToker would be subjected to comments.

Since most of the streamers do not involve moderators, sometimes the comments on the creators' physical features can enact a dark turn, specifically via the hyper-sexualization of underage girls, body shaming and inappropriate requests (e.g., "Can I masturbate?" or "Show me your feet"). When this occurs, the audience and the creator respond by insulting and subsequently banning any problematic commenters, in self-defence. Notably, most of the time these comments are not addressed by the creators.

The choice to remain silent can moreover represent an escape strategy from other undesired questions; most commonly this occurs when streamers do not disclose their age. It should be noted, in this sense, that nearly half of the TikTokers were younger than 16 years old. However, creators are faced with the necessity to keep this information confidential not because of its private nature, but as it represents an infraction of the platform's guidelines. In fact, in almost every video analysed, the creators included some personal data in their bio or while chatting (e.g., generalities, location or Instagram handle).

Tricking the platform by declaring a different age constitutes an unethical but easy way to gain access to the LIVE feature. A small part of the creators managed to trick the algorithm in a more complicated way, somehow figuring out how to stream even with fewer than 1000 followers. In addition, the research revealed a couple of other tendencies that demonstrate an awareness of the algorithm's functioning, namely: (1) stating that TikTok will shadow ban their content because of inappropriate behaviour (e.g., using curse words) and (2) avoiding filming an inappropriate action (such as smoking) by keeping it off screen.

In other words, while we have seen that users continuously acquire content creation skills by just performing LIVEs, the social interaction skills seem to be not so relevant—as demonstrated by the predominance of silence or the absence of a real moderation within the LIVEs. However, the attention to avoiding inappropriate behaviours or sharing sensitive information demonstrates the presence of online ethical behaviour skills, as users comprehend the potential risks of being exposed to a wider and unknown public.

Conclusions

While TikTok can tailor its users' experience in a very precise way without requiring them to express any personalization options, the three case studies detailed here demonstrate that online behaviour and social norms are the results of a negotiation between the users' grasp of TikTok's algorithm and the platform's specific affordances. On one hand, then, as Bucher and Helmond (2018) note, "How people behave, move, or simply exist in an environment affords important cues as to how others should behave, move or co-exist" (p. 9); on the other, TikTok's design and algorithm overlap in the ways they shape specific practices. It must be noted that the platform brands itself as a space in which users should "make every second count," focusing on the creation of appealing content and not on connections with peers and family. User-friendly editing tools, a regime of pseudonymity, and algorithmic ranking (Krutrök 2021) that pushes content towards a wider audience of strangers collectively reinforce elements of this process.

Taking into account the analysis undertaken to answer RQ1, it is possible to deduce that every practice or trick employed by TikTok teens sought to manage visibility. In this sense, it is possible to conclude that three main ways of applying algorithmic competency were detected: censorship, commodification and affordance exploitation. Censorship includes (1) the non-disclosure of information, such as age, in order to avoid sanctions by TikTok; (2) the modification of words, which may represent a form of "social steganography" (Boyd and Marwick 2011), though this masking of meaning is not motivated by the presence of acquaintances but rather by the perception of TikTok's "eyes"; (3) hiding behaviours that the platform discourages "in plain sight." Whatever the specific strategy, it appears clear that these practices represent a full awareness of algorithmic control.

With regard to commodification, the researchers noted three fundamental elements: (1) participating in or creating codified performances; (2) presenting the content as a means for acknowledgement in terms of economic reward and relatability; (3) the relevance of a curated appearance, in response to TikTok's beauty stigma (Melonio 2020). Commodification, in conclusion, means shaping content and self-presentation accordingly for TikTok's attention economy.

Lastly, with regard to the technical features of the platform, the exploitation of affordances is visible in: (1) the deployment of selected sounds; (2) the use of editing tools; (3) the way interactions are channelled in order to engage with a wider public. These pieces of evidence indicate a strategic use of a set of possibilities, ranging from structural features to the employment of the audience as part of the cog in the visibility mechanism (Zurovac and Boccia Artieri 2019).

These considerations allow us to understand the ways users can interact with and trick the algorithm, but also the digital skills involved in producing content on the platform (RQ2). As mentioned, this refers therefore to the evaluation of social interaction skills, content creation skills and ethical behaviour online, and of the relevance of each category.

Considering that the platform has low barriers to expression, thanks to its tools for content editing, users are not required to develop particular skills to produce appealing videos. Even when it comes to defining a "script," most of the time this is already codified by other users in the so-called trends from which a creator can choose, since participation through repetition is one of the main ways in which content is produced. In this sense, social interaction—as part of digital skills—seems to be the basis of the content production process. As noted by Abidin (2021b), this can assume the shape of "silosociality," although social interaction is also the context in which users learn how to perform by just existing. However, despite the centrality of interactions from the perspective of visibility, social interaction skills do not always seem to be particularly developed. Indeed, the three cases analysed here show different ways of interacting with others albeit with a shared purpose, as characterized by: (1) the lack of conversations happening in the comment section, (2) the presence of comments just for the sake of visibility; (3) the absence of norms regulating conversational exchanges—because even in the presence of moderators the interactions appeared to be mostly inappropriate, in terms of netiquette. This last observation is in line with what has been illustrated in the analysis regarding unethical behaviours. In fact, sustained by pseudonymity and the possible hugeness of an unknown audience, users tend to freely share content while perceiving the TikTok universe as something detached from their everyday life contexts.

That means that users are willing to share private or problematic content because what is licit is decided by the algorithmic imaginary and silosociality, under the false myth that what happens on TikTok stays on TikTok.

In conclusion, while it is not possible to assert whether or not the digital skills employed have been acquired before the encounter with the platform, these tales suggest that the algorithm (as it is perceived and imagined by the users) plays a significant role in making users develop peculiar practices (that we interpreted as algorithmic skills) in order to manage visibility.

References

Abidin C (2021a) From "networked publics" to "refracted publics": a companion framework for researching "below the radar" studies. Soc Media Soc 7:1–13. https://doi.org/10.1177/2056305120984458

Abidin C (2021b) Mapping internet celebrity on TikTok: exploring attention economies and visibility labours. Cult Sci J 12(1):77–103. https://doi.org/10.5334/csci.140

Alperstein N (2021) The present and future of performing media activism. In: Performing media activism in the digital age. Palgrave Macmillan, Cham, pp 211–235
Barradale G (2021) This is the story behind the 'step on the gas' TikTok audio. Thetab.com, June 16. Retrieved on https://thetab.com/uk/2021/06/16/this-is-the-story-behind-the-step-on-the-gas-tiktok-audio-210543
Bhandari A, Bimo S (2020) TikTok and the "algorithmized self": a new model of online interaction. AoIR Select Pap Internet Res. https://doi.org/10.5210/spir.v2020i0.11172
Bishop S (2019) Managing visibility on YouTube through algorithmic gossip. New Media Soc 21(11–12):2589–2606. https://doi.org/10.1177/1461444819854731
Boccia Artieri G (2020) Fare Sociologia attraverso l'algoritmo: potere, cultura e agency
Boccia Artieri G (2021) Networked participation: selfie protest and ephemeral public spheres. In: Della Ratta D, Lovink G, Numerico T, Sarram P (eds) The aesthetics and politics of the online self. ISBN: 978-3-030-65497-9
Boyd D, Marwick AE (2011) Social privacy in networked publics: teens' attitudes, practices, and strategies. In: A decade in internet time: symposium on the dynamics of the internet and society, September
Bucher T (2017) The algorithmic imaginary: exploring the ordinary affects of Facebook algorithms. Inf Commun Soc 20(1):30–44
Bucher T, Helmond A (2018) The affordances of social media platforms. In: Burgess J, Marwick A, Poell T (eds) The SAGE handbook of social media. SAGE
Chung AW (2020) Subverting the algorithm: examining anti-algorithmic tactics on social media. Doctoral dissertation, Massachusetts Institute of Technology
Eslami M, Karahalios K, Sandvig C, Vaccaro K, Rickman A, Hamilton K, Kirlik A (2016) First I "like" it, then I hide it: folk theories of social feeds. In: Proceedings of the 2016 CHI conference on human factors in computing systems, pp 2371–2382
Francisco MEZ, Ruhela S (2021) Investigating TikTok as an AI user platform. In: 2021 2nd international conference on computation, automation and knowledge management (ICCAKM), January. IEEE, pp 293–298
Gillespie T (2014) The relevance of algorithms. In: Media technologies: essays on communication, materiality, and society. MIT Press Scholarship, pp 167–193
Haddon L, Cino D, Doyle M-A, Livingstone S, Mascheroni G, Stoilova M (2020) Children's and young people's digital skills: a systematic evidence review. ySKILLS, Leuven
Haider J, Sundin O (2021) Information literacy as a site for anticipation: temporal tactics for infrastructural meaning-making and algo-rhythm awareness. J Doc. https://doi.org/10.1108/JD-11-2020-0204
Harriger JA, Evans JA, Thompson JK, Tylka TL (2022) The dangers of the rabbit hole: reflections on social media as a portal into a distorted world of edited bodies and eating disorder risk and the role of algorithms. Body Image 41:292–297
Harris M (2020) TikTok apologized for the glitch affecting the 'black lives matter' hashtag after accusations of censorship: 'we know this came at a painful time'. Insider, June 1. Retrieved on https://bit.ly/3fzehzb
Hill AL (2019) 'GRWM': modes of aesthetic observance, surveillance, and subversion on YouTube. In: Digitalität und Privatheit. pp 329–352
Hine C (2015) Ethnography for the internet: embedded, embodied and everyday. Bloomsbury, London
Karizat N, Delmonaco D, Eslami M, Andalibi N (2021) Algorithmic folk theories and identity: how TikTok users co-produce knowledge of identity and engage in algorithmic resistance. Proc ACM Human Comput Interact 5(CSCW2):1–44
Klug D, Qin Y, Evans M, Kaufman G (2021) Trick and please. A mixed-method study on user assumptions about the TikTok algorithm. In: 13th ACM web science conference 2021, June, pp 84–92
Krutrök ME (2021) Algorithmic closeness in mourning: vernaculars of the hashtag #grief on TikTok. Soc Media Soc 7(3):20563051211042396

Martin R (2019) TikTok, the Internet's hottest meme breeding ground, turns 1. NPR, August 5. Retrived on https://www.npr.org/2019/08/05/748163919/TikTok-the-internets-hottest-meme-breeding-ground-turns-1
Marwick AE, Boyd D (2011) I tweet honestly, I tweet passionately: twitter users, context collapse, and the imagined audience. New Media Soc 13(1):114–133
Medina Serrano JC, Papakyriakopoulos O, Hegelich S (2020) Dancing to the partisan beat: a first analysis of political communication on TikTok. In: 12th ACM conference on web science, July, pp 257–266
Melonio P (2020) TikTok's non-inclusive beauty algorithm & why we should care. IPHS 200: programming humanity. Paper 22
Noble S (2018) Algorithms of oppression: how search engines reinforce racism. NYU Press, New York
O'Neil C (2016) Weapons of math destruction: how big data increases inequality and threatens democracy. New York: Crown Publishers
Radovanović D, Hogan B, Lalić D (2015) Overcoming digital divides in higher education: digital literacy beyond Facebook. New Media Soc 17(10):1733–1749. https://doi.org/10.1177/1461444815588323
Savic M, Albury K (2019) Most adults have never heard of TikTok. That's by design. The Conversation, July 11. Retrieved on https://theconversation.com/most-adults-have-never-heard-of-TikTok-thats-by-design-119815
Schellewald A (2021) On getting carried away by the TikTok algorithm. AoIR Select Pap Internet Res
Shifman L (2013) Memes in a digital world: reconciling with a conceptual troublemaker. J Comput-Mediat Commun 18(3):362–377
Smith B (2021) How TikTok reads your mind. Retrieved from https://www.nytimes.com/2021/12/05/business/media/tiktok-algorithm.html
Statista (2021) Distribution of global TikTok creators 2021, by age group. Retrieved on https://www.statista.com/statistics/1257721/TikTok-creators-by-age-worldwide/
TikTok (2020) How TikTok recommends videos #ForYou, June 18. https://newsroom.TikTok.com/en-us/how-TikTok-recommends-videos-for-you/
TikTok (2021) All the ways you can enjoy LIVE with TikTok. https://newsroom.TikTok.com/en-us/TikTok-live-features-2021
Van Dijck J (2013) Social media platforms as producers. In: Producing the internet. Critical perspectives of social media. Nordicom, pp 45–62
Vázquez-Herrero J, Negreira-Rey M-C, López-García X (2020) Let's dance the news! How the news media are adapting to the logic of TikTok. Journalism. https://doi.org/10.1177/1464884920969092
Wall Street Journal (2021) Investigation: how TikTok's algorithm figures out your deepest desires. Retrieved on https://www.wsj.com/video/series/inside-TikToks-highly-secretive-algorithm/investigation-how-TikTok-algorithm-figures-out-your-deepest-desires
Weimann G, Masri N (2020) Research note: spreading hate on TikTok. Stud Conflict Terror. https://doi.org/10.1080/1057610X.2020.1780027
Wright L (2021) The sound of identity: audios and hashtags as nexuses of practice on TikTok
Yu JX (2019) Research on TikTok APP based on user-centric theory. Appl Sci Innov Res 3(1):28
Zulli D, Zulli DJ (2020) Extending the internet meme: conceptualizing technological mimesis and imitation publics on the TikTok platform. New Media Soc 24:1461444820983603
Zurovac E, Boccia Artieri G (2019) Performing and perceiving the microcelebrity status in snapchat: an Italian case study. In: Desecrating celebrity. Proceedings of the IV international celebrity studies journal conference. https://doi.org/10.4458/2719

Part III
Digital Literacy and Communities of Practice

Chapter 9
Digital Literacy and Agricultural Extension in the Global South

Gordon A. Gow, Uvasara Dissanayeke, Ataharul Chowdhury, and Jeet Ramjattan

Introduction

The global agriculture sector is undergoing profound changes because of digital transformation or digitalization (Matos et al. 2020). While the impact of COVID-19 has helped to accelerate the digitalization process, the disruption caused by the pandemic has also revealed systemic barriers in parts of the world where digital literacy remains relatively low (Ceballos et al. 2020; Mohapatra 2020). So far, research into the social and economic impact of digital transformation of the agriculture sector has come mainly from researchers working in the Global North (Bronson and Knezevic 2019; Phillips et al. 2019). However, the UN's Food and Agriculture Organization (FAO) and the World Bank are now encouraging digital agriculture in the Global South and there is growing recognition that 'a critical approach toward the pervasive application of digital technologies in developing and emerging country agriculture is much needed' (Klerkx 2019, p. 12).

G. A. Gow (✉)
Department of Sociology and Media & Technology Studies, University of Alberta, Edmonton, Alberta, Canada
e-mail: ggow@ualberta.ca

U. Dissanayeke
Department of Agricultural Extension, University of Peradeniya, Kandy, Sri Lanka

A. Chowdhury
School of Environmental Design and Rural Development, University of Guelph, Guelph, ON, Canada

J. Ramjattan
Department of Agricultural Economics and Extension, The University of the West Indies, Saint Augustine, Trinidad

Extension Training and Information Services Division, Ministry of Agriculture, Land and Fisheries, San Fernando, Trinidad and Tobago

D. Radovanović (ed.), *Digital Literacy and Inclusion*,
https://doi.org/10.1007/978-3-031-30808-6_9

Small-scale farmers are particularly exposed to unintended negative impacts of digitalization (Ending hunger 2020; Pereira et al. 2018). Smallholders produce more than 70% of the food consumed in countries of the Global South (Fanzo 2017; FAO 2020), where they play a crucial role in maintaining the genetic diversity of the food supply and contribute to food security for many cities worldwide, playing a vital role in the UN's Sustainable Development Goals (Lowder et al. 2016). These are usually family-operated farms with small plots of land located in rural areas and traditionally low-tech operations. However, researchers and activists are concerned that digitalization efforts in the Global South will tend to marginalize the voices of smallholders while altering established farming practices and patterns of social relations in their communities, especially for women and youth (Bronson 2018; Fraser 2019; GRAIN 2021; Rotz et al. 2019).

Despite these concerns, there is evidence that digitalization can also help farmers maintain their independence and introduce innovative practices to enhance their livelihoods (Cisneros and Roberts 2021; Matthews 2017). An important step in this direction is to empower smallholders to participate more actively in discussions and decisions related to digital agriculture (Bonina et al. 2021). Agriculture extension and advisory services (EAS) can contribute to these efforts by supporting digital literacy development within these communities (Dlamini and Worth 2019; Shilomboleni et al. 2020; Steinke et al. 2020). Laurens Klerkx, a leading voice in this area, states that 'what is crucial to acknowledge is that there is a plurality of [digital] transition pathways which co-exist, intersect, collaborate, or compete' (Klerkx 2020, p. 132). EAS organizations will play a crucial role in guiding those transitions in their work with smallholders and rural communities. Moreover, digital literacy is now considered central to a 'new extensionist' agenda and national 'e-agriculture' strategies (Davis 2015; Ganpat et al. 2016; Wanigasundera and Atapattu 2019).

EAS organizations will need to look beyond the immediate training needs of their field staff (Narine and Harder 2019; Norton and Alwang 2020) to introduce digitalization strategies that take into account the priorities, aspirations, and constraints of smallholders and the communities they serve. The ability of smallholders to collectively assert community-based control and autonomy in decisions regarding digital practices, as well as the data produced by those practices, is aligned with the concept of digital self-determination (Remolina and Findlay 2021, p. 18). Advancing the prospects of digital self-determination for smallholders raises three important questions: What is the role of EAS organizations when it comes to promoting digitalization in agriculture? How does digital literacy figure into this effort? What practical steps can EAS practitioners take to empower smallholders to make reasoned choices regarding digital ICTs and their integration into local work practices and processes?

The intent of this chapter is to begin to address these questions by presenting a conceptual framework that draws a connection between the literature on agriculture innovation systems with an interactionist view of digital literacy from organizational studies. We then explain how this view of digital literacy aligns with a capabilities-centric approach within ICT for development (ICT4D) by situating it along four degrees of empowerment from Kleine's Choice Framework (Kleine

2013). The final section of the chapter includes examples of how this framework can be used, which are based on preliminary findings from an action research study involving EAS practitioners in Trinidad and Sri Lanka (Gow et al. 2020b). The overall goal of that ongoing research project is to improve our understanding of the relationship between EAS practitioners, digital capabilities, and inclusive innovation within the agriculture sector, especially among smallholders and their communities.

Agricultural Extension: A Brief Overview

The origins of agricultural extension and advisory services (EAS) in the Global South began with efforts by colonial governments during the nineteenth century to improve crop yields in agricultural products for export. With the advent of the Green Revolution in the 1950s and 1960s, these countries were encouraged to introduce national agricultural advisory services, focusing on transferring knowledge and skills from research institutions to farmers and farming communities as a strategy to apply modern science to crop production (Ganpat 2013). For much of its history, EAS practitioners applied a 'linear model' of technology transfer within a modernization paradigm of development that sought to introduce innovations in agriculture technology primarily from the Global North with the assistance of government extension officers working directly with farmers (Heeks 2018).

Today, the linear transfer model co-exists with an agricultural innovation systems (AIS) framework 'focusing more broadly on the factors that stimulate innovative behavior and stress[ing] linkages and partnerships with a wide range of actors along agricultural value chains, including the agribusiness sector' (Anderson 2008, p. 9). While the AIS paradigm has yet to be fully embraced by EAS organizations in the Global South, it reimagines the agricultural extension system as a bridging institution that coordinates the flow of information across multi-stakeholder networks involving farmers and other actors such as agriculture scientists and educators, suppliers of inputs and financial services, as well as consumer-facing stakeholders responsible for food safety, distribution, and marketing.

Another significant change over the previous two decades has been a sharp decline in funding of public sector EAS accompanied by other structural changes that have given rise to a plurality of alternative providers that include the private sector, non-profit, and producer organizations in this domain (Benson and Jafry 2013; Blum et al. 2020). Nonetheless, a vital role for public sector EAS continues within the AIS paradigm because it often serves smallholder communities who typically have a complex portfolio of needs, a limited ability to pay for private services, and who could benefit from a diversification of livelihood strategies to support vulnerable and marginal groups, such as women and the poor (Benson and Jafry 2013, p. 389). In meeting this need, public sector EAS is uniquely positioned to lead digitalization efforts with smallholders and the rural communities in which they work and live.

Digital Capabilities in the Agriculture Sector

Research on digital transformation of the agriculture sector has identified several factors that influence farmers' adoption and use of technology. Khanna (2021) suggests a distinction between 'technological factors' and 'farmer characteristics' among a range of considerations in understanding adoption patterns. Within the category of technological factors, learning barriers are significant when farmers encounter unfamiliar and often technically complex equipment and services. Under these conditions, farmers seek information from others within their community, with EAS practitioners playing an instrumental role 'critical in building trust in the technology, lowering learning costs, and protecting farmer interests' (Khanna 2021, p. 1233). The uncertainty surrounding digital transformation will also prompt anxieties about deskilling, data ownership, and privacy as farmers confront technologies requiring them to delegate important decisions to machine learning systems and share information with third-party service providers about their cultivation practices, input use, and yields.

Behavioural factors also play a role in shaping digital transformation efforts. Farmer's adoption decisions are affected by a range of individual cognitive processes, such as perceptions about the risk to their livelihoods using cost-benefit assessments based on local conditions, as well as emotional disposition and personal experience. These behavioural factors extend to include the influence of group dynamics when introducing farmers to new technology-related agricultural practices during interactions with peers and other influential actors (Khanna 2021, p. 1234).

Digital literacy levels among farmers in the Global South have been examined in recent studies (Khan et al. 2020), with consistent findings showing a positive correlation between education levels, digital skills, and adoption of new technologies. However, the UN Food and Agriculture Organization (FAO) has identified low 'e-literacy and digital skills' as barriers to ICT adoption in the agricultural and rural development sector. In these settings, deficiencies in basic literacy, numeracy, and access to computing courses, further limit digital skills attainment (FAO 2019, p. 4). Nonetheless, policymakers and practitioners view it as essential for digital transformation:

> Digital skills and e-literacy remain a significant constraint to the use of new technologies and are particularly lacking in rural areas, especially in developing countries. The diversity of available digital technologies and a lack of standardisation also present a barrier to adoption. The choice of which technology to use is complex and there is a lack of advisory services to support farmers in these decisions. Education and supporting services must be improved to support the adoption of digital technologies. (FAO 2019, p. 15)

Despite its frequent mention, many of these studies and reports devote little attention to critically reflecting on digital literacy and how it fosters capabilities that can promote pathways towards digital self-determination among smallholders within the wider agricultural innovation system.

An Interactionist View of Digital Literacy

A lack of critical reflection on the concept of digital literacy is not unique to the agriculture sector but is also found in organizational studies on digital transformation (Cetindamar Kozanoglu and Abedin 2021). While that literature has identified digital literacy as a 'critical dynamic capability of organizations during their digital transformations,' it has tended to focus on individuals rather than the social context of technology adoption (Cetindamar Kozanoglu and Abedin 2021, p. 1650). This realization has prompted efforts to conceptualize an 'interactionist approach' to digital literacy research.

The interactionist approach introduces the notion of *organizational affordances* based on recent developments in affordance theory and its application to the study of information systems (Cetindamar Kozanoglu and Abedin 2021). Affordance theory first appeared in the work of James Gibson, the founder of ecological psychology, and was later incorporated into human–computer interaction (HCI) research by Donald Norman (Anderson and Robey 2017). In its early conception, the theory tended toward a cognitivist view in understanding how individuals came to interpret and use technologies. Good design practice was considered essential in revealing to users the intrinsic and intended features and functions—*the affordances*—of an artefact or software application. Drawing on adaptive structuration theory and Orlikowki's 'practice lens' for studying technology in organizational settings (Orlikowski 2000), the concept of affordance has since been expanded to distinguish between 'affordances in information' and 'affordances in articulation' to describe differences between the features incorporated into design versus the social context of use (Vyas et al. 2016). These two types of affordances are combined into an 'organizational affordances' model that accounts for the mutual influence of individual and group level dynamics on technology adoption in the workplace.

In applying the organizational affordances model to digital literacy, Cetindamar Kozanoglu and Abedin (2021) reference Stordy's taxonomy of literacies (Stordy 2015), making a crucial distinction between *autonomous* and *ideological* models of literacy. Whereas autonomous models tend to view literacies as 'an individual's intellectual abilities … for which they are largely responsible,' ideological models by contrast 'view literacy as a social practice that cannot be detached from its context which both creates and perpetuates it' (Stordy 2015, p. 460). Stordy's analysis further suggests that training programmes focussed on autonomous literacy tend to promote conformity as a pathway to self-improvement and emphasize workplace productivity as a primary objective. On the other hand, ideological approaches tend to align with a holistic human development paradigm that views literacy training as a foundation for critical thinking, empowerment, and community building. Stordy synthesizes these two views to form a definition of literacy as an individual cognitive skill within the context of group action:

> (t)he abilities a person or social group draws upon when interacting with digital technologies to derive or produce meaning and the social learning and work practices that these abilities are applied to. (Stordy 2015, p. 472)

This definition provides the basis for an interactionist model of digital literacy that encompasses 'affordances in information' and 'affordances in articulation' (Cetindamar Kozanoglu and Abedin 2021). Digital literacy in relation to *information affordances* refers to an individual's understandings and interpretations of a technology or software application. These might be considered the 'what' aspects of a technology evident to a user (Vyas et al. 2006). Novice users will presumably have more modest abilities concerning the 'what' aspects than experienced users. For example, an individual's competence in utilizing the features and functions of a word-processing application will expand as he or she is exposed to and trained in its use. As such, the actualization of information affordances—or what Anderson and Robey (2017) refer to as 'affordance potency'—will be closely related to an individual's exposure to a device and/or software application. As Norman's earlier work sought to demonstrate, good design is integral to this actualization, but information affordances also extend to encompass the role of formal training, exposure to marketing materials, and informal learning that happens when individuals on their own explore the features of a technology.

Digital literacy in relation to *articulation affordances* refers to a shared set of procedural understandings of a technology in use (Vyas et al. 2006, p. 95). These are the 'how to' aspects that emerge as technology is enacted in practice or the 'this-is-how-we-do-it-here' dimension established and reinforced in the context of group dynamics or professional practice. Articulation affordances are expressions of the adoption of technology in specific social settings. For example, in some contexts, such as large commercial publishing, a word processing application may be integrated into a business process that actualizes many of the advanced features and functions of the software during manuscript preparation. Compare that example with a small group of volunteers that requires only the most basic features of the same software to produce a community newsletter. Each group uses the same technology, but the varying social contexts of use actualize different articulation affordances.

These examples illustrate Stordy's definition of digital literacy as an *interaction* between individual skills relative to the context of use. This dialectical relationship is expressed in an 'organizational affordances' model to guide digital literacy development within the workplace (Cetindamar Kozanoglu and Abedin 2021). Figure 9.1

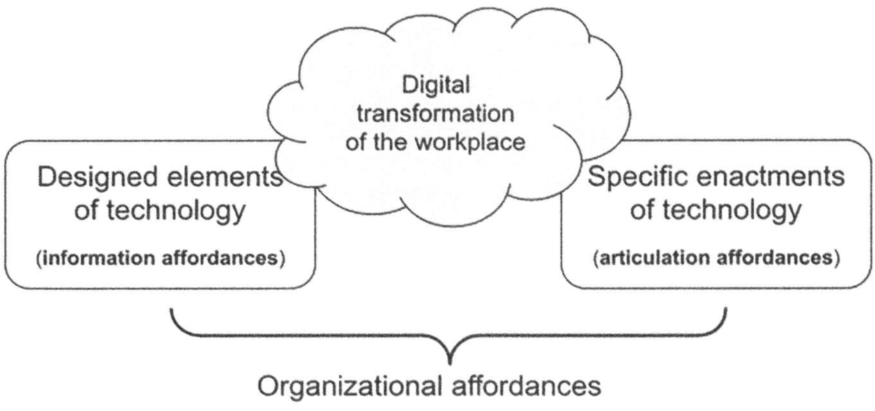

Fig. 9.1 The organizational affordances model adapted from (Vyas et al. 2016). Digital transformation encompasses both information and articulation affordances

depicts this model to illustrate how the digitalization outcomes emerge through the ongoing interaction of information and articulation affordances.

EAS Organizations and the Digital Literacy Dilemma

The organizational affordances model presents a dilemma for EAS organizations when it comes to planning a strategy for promoting digital literacy. Emphasis on digital skills in relation to specific technologies—an 'information affordances' focus—will be necessary to establish agricultural workers competent in using those technologies. However, this by itself might not create sufficient conditions for workers to adopt a new digital practice. While farmers may perceive value in such training, other priorities may take precedent when it comes to their decisions about using new ICTs. Getting fair prices for their crops, calling for improvements in infrastructures to transport their produce to markets, as well as other community concerns will be more immediate concerns for farmers (Iazzolino 2021). In other words, despite efforts at providing them with digital skills training, farmers and other workers may remain reluctant to use new technologies because they are considered a distraction from more pressing concerns or because others in the community are not yet using them.

One response may be for organizations to simply go ahead and launch a digital initiative and impose on agriculture workers to adapt accordingly. For example, farmers might be required by law to adopt specific digital practices to conform to a food traceability system introduced by a government department or a large buyer. This is a top-down initiative that actualizes a prescribed set of articulation affordances with digital ICTs. Such a strategy runs the risk of overlooking local conditions and disrupting established processes on the farm and relationships across the agricultural value chain. For example, the Indian government's controversial Agristack initiative will require all farmers to conform to a 'Unified Farmer Service Interface' designed by Microsoft as mandated by the national government (Kapil 2021). However, opponents of Agristack (Internet Freedom Foundation 2021) have argued that this policy threatens to disempower smallholders while undermining long-standing practices, thereby limiting future opportunities for inclusive digitalization efforts within the agriculture innovation system.

Resolving the digital literacy dilemma for smallholder farmers in the Global South will require EAS organizations to incorporate a multi-faceted approach that actualizes both information and articulation affordances as they introduce farmers and other agriculture workers to new digital practices. On the one hand, training will need to focus on developing individual digital skills with respect to the information affordances of specific ICT tools and systems. These will include competencies such as information and data literacy, communication and collaboration, digital content creation, online safety, and problem-solving with ICTs. On the other hand, training will also need to factor in articulation affordances by considering how ICTs will be integrated into and transform established social practices among farmers and other agriculture workers. For example, the ability of smallholders to actualize the features and functions of a digital crop monitoring and management system may be limited as compared with agribusiness operations that have dedicated resources for training and IT support at their disposal.

A digitalization strategy targeted to smallholders can benefit from an interactionist approach to digital literacy because it recognizes a dialectical relationship between individual skills and group-level dynamics in the adoption and use of ICTs. The organizational affordances model presents digitalization as an emergent outcome resulting from the interaction of information affordances and articulation affordances, suggesting that both aspects will need to be considered when it comes to planning and evaluating a digital literacy training programme.

Putting It into Practice: The Technology Stewardship Training Programme

This section illustrates how an interactionist approach to digital literacy has been incorporated into an ICT training programme for agricultural extension officers in the Global South. We draw on initial results from qualitative research conducted with EAS practitioners, highlighting examples from Sri Lanka and Trinidad. The project methodology is similar to the Ethnographic Action Research (EAR) design for ICT4D projects first introduced by Tacchi and her colleagues in various Southeast Asian countries, including Sri Lanka (Tacchi 2015). Our project focusses on EAS practitioners, who are trained in a set of ICT4D-related skills and then invited to become co-researchers serving as liaisons between the communities they serve and the academic research team.

The training programme also builds on previous efforts to study agricultural workers as communities of practice (Adelle et al. 2021; Morgan 2011; Nuutinen and Filho 2018; Tran et al. 2018; Triste et al. 2018). More specifically, our project focusses on 'technology stewardship' as a catalyst for digitalization. Wenger et al. (2009) introduce the term technology stewardship to describe a role for individuals who support the decisions to select and use digital technologies within a community of practice (CoP). Technology stewardship is an informal leadership role for cultivating the 'digital habitat' with the members of a CoP:

> Technology stewards are people with enough experience of the working of a community to understand its technology needs and enough experience with or interest in technology to take leadership in addressing those needs. Stewarding typically includes selecting and configuring the technology and supporting its use in the practice of the community. (Wenger et al. 2009, p. 25)

EAS practitioners are good candidates for this role because they typically represent knowledgeable intermediaries within one or more communities of practice. Research findings from other ICT4D studies suggest the type of intermediary role played by EAS practitioners is influential in fostering inclusivity within the agriculture innovation system (Heeks 2018, p. 60), particularly when it comes to smallholders as well as women and other marginalized groups living in rural communities (Ayre et al. 2019; Oreglia 2014; Walsham 2020).

In our adaptation, the intermediary role of the technology steward is guided by Kleine's four-step empowerment model or 'Choice Framework' (Kleine 2013). The Choice Framework sets out a progression of capacity building activities from a basic introduction to ICT and leading up to its integration into practice, which involves four overlapping stages (Kleine 2010). The first stage begins with creating an *awareness* of technology choices that are available to a community of practice (CoP). The next stage is then to foster *a sense of choice* by providing examples of how members of the CoP might apply technology to address an existing priority or aspirational objective. The third stage involves facilitating the *use of choice* through pilots or prototypes with specific ICT solutions. Kleine refers to the fourth stage as the *achievement of choice* to indicate a transformative moment when members of the CoP, guided by the technology steward, assess the suitability of the chosen ICT solution with respect to community needs and ambitions. Normatively speaking, the primary development aim is therefore to promote 'choice itself' (Kleine 2011, p. 125) as an essential step towards digital self-determination.[1]

Our project has drawn from these various sources to create and introduce a training programme titled 'ICT Stewardship for Agricultural Communities of Practice' that was tested with EAS practitioners between 2016 and 2019, involving two cohorts in Sri Lanka in partnership with the University of Peradeniya, and two cohorts in Trinidad in partnership with the University of the West Indies. To date, a total of 80 EAS practitioners have participated in the programme, and future offerings are now being planned. Starting in 2018, cohort members were invited to complete a capping project by conducting a small-scale action research project or 'campaign' with a community of practice of their choosing. These capping project campaigns provide an opportunity to conduct participatory action research with EAS practitioners as they take up an intermediary role intended to foster 'situated learning' (Lave and Wenger 1990) between farmers, other community members, technology sponsors, and academic researchers.

Figure 9.2 depicts the overall project design, showing the relationship between technology stewardship training and the capping project campaign as a form of action research. The dialectic between information and articulation affordances that arises out of the action research campaign fosters emerging capabilities along the four stages of empowerment in the Choice Framework. These capabilities become the basis on which a CoP can begin to make reasoned choices about its pathway to digitalization.

[1] We would note that Kleine's Choice Framework and its normative orientation toward capabilities aligns with the Stordy's 'ideological model' that views literacy less as a vehicle for conformity and more as a catalyst for critical thinking and empowerment.

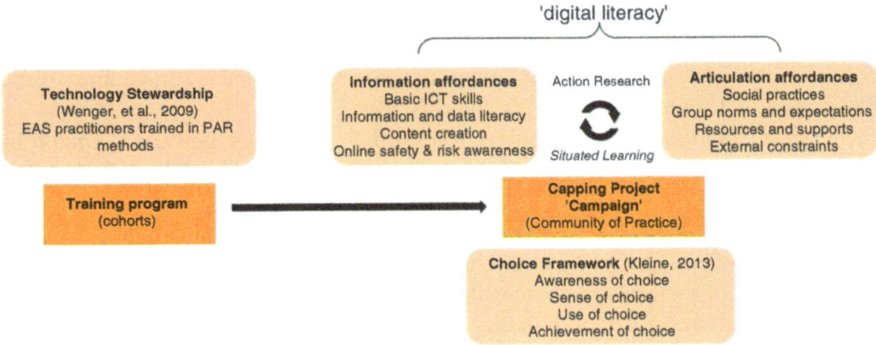

Fig. 9.2 The technology stewardship programme as an interactionist approach to digital literacy

Insights from Two Capping Project Campaigns

Elsewhere we have provided specific details on our research design, implementation, and initial results from the pilot phase of the technology stewardship project (Gow et al. 2020a, c). For the purpose of this chapter, we will consider two capping project reports that illustrate the value of an interactionist model in guiding the efforts of a technology steward as they lead digitalization efforts with a community of practice.

From Articulation Affordances to Information Affordances—Assessing Individual Digital Skills

Technology stewards are trained to create a campaign goal statement that includes a specific target for a specific activity, and with a clearly defined community of practice in mind. The stewards are instructed not to identify a particular digital tool or platform when creating the goal statement but instead to work with the community to describe a 'communication action' priority derived from a set of 'community orientations' set out in Wenger et al. (2009, p. 69). In effect, the campaign goal statement is intended to emphasize articulation affordances by identifying deficiencies in social practices or organizational processes that might be addressed with the application of ICT.

The campaign goal statement provides a basis on which the technology steward can then identify the information affordances that will be required in an ICT solution. Having identified a set of information affordances, the technology steward can then undertake a digital skills assessment of community members. For example, Fig. 9.2 shows a planning table created by Suranjan, an instructor with the Sri Lanka

Department of Agriculture, who attended our training course in 2018. His capping project identified 'Seed paddy producers in the Galle District' as the community of practice. The communication action in the campaign goal statement is 'improving access to expertise,' which led them to focus on articulation affordances and to think differently about how they used their mobile devices when trying to reach EAS officers for advice. This approach increased the awareness and sense of choice among the farmer group as they realized that asynchronous messaging could serve as an alternative practice to relieve some of the problems associated with voice calls. Suranjan then conducted a series of activities that focussed on the information affordances of mobile phones in relation to asynchronous messaging (Fig. 9.3). In taking this step he could assess the individual competencies of the farmers with their phones and was then able to determine what type of digital skills training would be needed prior to piloting a text messaging campaign with this group.

Activity	**Tool/feature**	**Configuration Notes**	**Test Results**
Farmers able to receive massages	Use normal SMS service facilitate by the particular sim service provider	Seed paddy produces in the Galle District western zone	Tested with their mobile phones - √ Eg: Nokia, Samsung, Huawei, Brandtel etc.
Farmers able to know and verifying the massage received from Technical steward	Use phone features	Choose specific massage received alert tone not common for all	Set the receiving tone and tested- √
Farmers able to see the sending massage in their phone screen	Use phone display	Shod be shorten and not include technical jargons. Not too long.	Single massage for single matter.- √ Not good for general answers avoid multiple answers. - √
Farmers able to clean the massage box when it filled	Use phone features to delete inbox familiarized them to massage warning display symbol - phone features	Practice Aware of them about this scenario	Tested with their mobile phones - √

Prototyping and Testing Worksheet

Fig. 9.3 Suranjan's 'Prototyping and Testing Worksheet' used to assess farmers' proficiency in using the information affordances of their mobile phones for asynchronous text messaging

From Information Affordances to Articulation Affordances—Revealing Group Preferences

A technology steward might also adopt a reverse strategy that begins with information affordances but then assesses them against articulation affordances. This can provide valuable insights when it comes to assessing the use and achievement of choice with respect to a specific ICT as taken up in practice. For example, one capping project report revealed how the information affordances of what seemed like an ideal ICT platform did not align well with existing preferences when introduced in a workplace setting.

Antoinette is a researcher and outreach coordinator with the Cocoa Research Centre (CRC) in Trinidad, who participated in the 2019 training at the University of the West Indies. The CRC provides extension services for local farmers and is part of a wider community of practice that includes members involved in cocoa research, production, and marketing. Her campaign attempted to introduce two digital messaging platforms in conjunction with the World Cocoa and Chocolate Day Expo. In a follow-up email with the research team, she described her insights from the experience:

> I think [communications for] the event could have been better managed with ICT, but our team is somewhat in need of convincing (let's say) with regard to the efficacy of it ... so I used it and mostly observed others reactions to whenever it was mentioned ... I think maybe a less complicated undertaking would be a better candidate for testing out an ICT with my colleagues ...

During the planning stage, she had identified the necessary information affordances of the chosen messaging platforms as being relevant to the campaign goal. However, her remarks 'I think maybe a less complicated undertaking would be a better candidate ...' indicate that her campaign revealed that group preferences for messaging did not align well with the designed features of the chosen platforms. In other words, the campaign revealed articulation affordances related to the group's preferred messaging practices that did not correspond with Antoinette's initial expectations. No matter how much effort she might have put into training and encouraging her colleagues to use the messaging tools, the communication practices and norms within the group appear not to have aligned well with the information affordances of the ICT tool. Antoinette's evaluation will assist in guiding her future efforts with her team as she now has a greater appreciation for the uncertain relationship between the available choice of ICTs and how these may be taken up to achieve new digital practices in the workplace.

Conclusion

EAS practitioners serve a crucial intermediary role in digitalization of the agricultural sector in the Global South. An interactionist approach to digital literacy will be important to consider as they work with communities of practice to facilitate new practices involving unfamiliar ICTs. Importantly, the organizational affordances model conceptualizes digital literacy as a dialectical relationship between individual cognitive skills and group-level practices with ICT. Initial results from our work with EAS practitioners in Sri Lanka and Trinidad illustrate how an interactionist approach to digital literacy can inform efforts by EAS organizations to develop training strategies that will include both individual digital literacy training and group-level assessment of digital practices.

Our ongoing research in this area will further introduce and evaluate practical techniques and methods for EAS practitioners to foster situated learning opportunities encompassing both information and articulation affordances as a strategy for digital capabilities development. Future research might also consider how an interactionist view on digital literacy can provide further insights into efforts to apply Kleine's Choice Framework in other settings and to capabilities-oriented ICT4D more generally.

Acknowledgements This project has been made possible with funding from the Social Sciences and Humanities Research Council of Canada, Global Affairs Canada Emerging Leadership in the Americas programme, and with generous support from the Postgraduate Institute of Agriculture at the University of Peradeniya and the Faculty of Food and Agriculture at the University of the West Indies. The author would like to thank Professor Wayne Ganpat from the Faculty of Food and Agriculture at the University of the West Indies for his ongoing support of the project. The authors also wish to express their appreciation to the cohorts from our technology stewardship training courses in Sri Lanka and Trinidad who contributed to this research through their active participation in classroom discussion and activities.

References

Adelle C, Kroll F, Losch B, Görgens T (2021) Fostering communities of practice for improved food democracy: experiences and learning from South Africa. Urban Agric Reg Food Syst 6(1):e20007. https://doi.org/10.1002/uar2.20007

Anderson JR (2008) Agricultural advisory services. World Bank, Washington, DC. https://openknowledge.worldbank.org/handle/10986/9041

Anderson C, Robey D (2017) Affordance potency: explaining the actualization of technology affordances. Inf Organ 27(2):100–115. https://doi.org/10.1016/j.infoandorg.2017.03.002

Ayre M, Mc Collum V, Waters W, Samson P, Curro A, Nettle R et al (2019) Supporting and practising digital innovation with advisers in smart farming. NJAS Wagening J Life Sci 90-91:100302. https://doi.org/10.1016/j.njas.2019.05.001

Benson A, Jafry T (2013) The state of agricultural extension: an overview and new caveats for the future. J Agric Educ Ext 19(4):381–393. https://doi.org/10.1080/1389224X.2013.808502

Blum ML, Cofini F, Sulaiman RV (2020) Agricultural extension in transition worldwide: policies and strategies for reform, Report no. FAO, Rome. https://doi.org/10.4060/ca8199en

Bonina C, Koskinen K, Eaton B, Gawer A (2021) Digital platforms for development: foundations and research agenda. Inf Syst J n/a(n/a). https://doi.org/10.1111/isj.12326

Bronson K (2018) Smart farming: including rights holders for responsible agricultural innovation. Technol Innov Manag Rev 8(2):7–14

Bronson K, Knezevic I (2019) The digital divide and how it matters for Canadian food system equity. Can J Commun 44(2). https://doi.org/10.22230/cjc.2019v44n2a3489

Ceballos F, Kannan S, Kramer B (2020) Impacts of a national lockdown on smallholder farmers' income and food security: empirical evidence from two states in India. World Dev 136:105069. https://doi.org/10.1016/j.worlddev.2020.105069
Cetindamar Kozanoglu D, Abedin B (2021) Understanding the role of employees in digital transformation: conceptualization of digital literacy of employees as a multi-dimensional organizational affordance. J Enterp Inf Manag 34(6):1649–1672. https://doi.org/10.1108/JEIM-01-2020-0010
Cisneros A, Roberts T (2021) Decolonizing innovation. Retrieved from https://bostonreview.net/forum_response/decolonizing-innovation/?utm_source=Digest&utm_campaign=8f1bf7851c-RSS_EMAIL_CAMPAIGN&utm_medium=email&utm_term=0_d90a01c7ff-8f1bf7851c-87816585
Davis K (2015) The new extensionist: core competencies for individuals. Global Forum for Rural Advisory Services. http://www.g-fras.org/en/knowledge/gfras-publications.html?download=358:the-new-extensionist-core-competencies-for-individuals
Dlamini MM, Worth S (2019) The potential and challenges of using ICT as a vehicle for rural communication as characterised by smallholder farmers. Asian J Agric Ext Econ Sociol 34(3):1–10. https://doi.org/10.9734/ajaees/2019/v34i330202
Ending hunger (2020) Ending hunger: science must stop neglecting smallholder farmers. Nature 586:336. https://doi.org/10.1038/d41586-020-02849-6
Fanzo J (2017) From big to small: the significance of smallholder farms in the global food system. Lancet Planet Health 1(1):e15–e16. https://doi.org/10.1016/S2542-5196(17)30011-6
FAO (2019) Digital technologies in agriculture and rural areas. Retrieved from http://www.fao.org/3/ca4985en/ca4985en.pdf
FAO (2020) Farm family knowledge platform: small-scale fisheries and aquaculture & family farming. Retrieved from http://www.fao.org/family-farming/home/en/
Fraser A (2019) Land grab/data grab: precision agriculture and its new horizons. J Peasant Stud 46(5):893–912. https://doi.org/10.1080/03066150.2017.1415887
Ganpat W (2013) The history of agricultural extension in Trinidad and Tobago. Randle, Kingston
Ganpat WG, Ramjattan J, Strong R (2016) Factors influencing self-efficacy and adoption of ICT dissemination tools by new extension officers. J Int Agric Ext Educ 23(1). https://doi.org/10.5191/jiaee.2016.23106
Gow G, Chowdhury A, Ramjattan J, Ganpat W (2020a) Fostering effective use of ICT in agricultural extension: participant responses to an inaugural technology stewardship training program in Trinidad. J Agric Educ Ext 26(4):335–350. https://doi.org/10.1080/1389224X.2020.1718720
Gow G, Dissanayeke U, Jayathilake C, Chowdhury A, Ramjattan J, Ganpat W, Rathnayake S (2020b) Putting the capabilities approach into action (Research): a comparative assessment of a technology stewardship training program for agricultural extension in Sri Lanka and Trinidad. Paper presented at the International Association for Media and Communication Researchers (IAMCR), Tampere
Gow G, Dissanayeke U, Jayathilake H, Kumarasinghe I, Ariyawanshe K, Rathnayake S (2020c) ICT leadership education for agricultural extension in Sri Lanka: assessing a technology stewardship training program. Int J Educ Dev Using Inf Commun Technol 16(1):35–43
GRAIN (2021) Digital control: how Big Tech moves into food and farming (and what it means). https://grain.org/en/article/6595-digital-control-how-big-tech-moves-into-food-and-farming-and-what-it-means
Heeks R (2018) Information and Communication Technology for Development (ICT4D). Routledge, London
Iazzolino G (2021) What about the crates? Rethinking digital farming in Kenya. Africa at LSE
Internet Freedom Foundation (2021) A thoroughly bad IDEA: our comments on the agristack consultation paper. Retrieved from https://internetfreedom.in/iff-response-to-the-idea-paper-on-agristack/

Kapil S (2021) Agristack: the new digital push in agriculture raises serious concerns. DownToEarth. Retrieved from https://www.downtoearth.org.in/news/agriculture/agristack-the-new-digital-push-in-agriculture-raises-serious-concerns-77613

Khan NA, Qijie G, Sertse SF, Nabi MN, Khan P (2020) Farmers' use of mobile phone-based farm advisory services in Punjab, Pakistan. Inf Dev 36(3):390–402. https://doi.org/10.1177/0266666919864126

Khanna M (2021) Digital transformation of the agricultural sector: pathways, drivers and policy implications. Appl Econ Perspect Policy 43(4):1221–1242. https://doi.org/10.1002/aepp.13103

Kleine D (2010) ICT4WHAT?—using the choice framework to operationalise the capability approach to development. J Int Dev 22(5):674–692. https://doi.org/10.1002/jid.1719

Kleine D (2011) The capability approach and the 'medium of choice': steps towards conceptualising information and communication technologies for development. Ethics Inf Technol 13(2):119–130. https://doi.org/10.1007/s10676-010-9251-5

Kleine D (2013) Technologies of choice? ICTs, development, and the capabilities approach. MIT Press, Cambridge

Klerkx L (2019) Social science on digitalization in agriculture– established and emerging strands of work and future avenues. Paper presented at the Séminaire DigitAg, Montpellier, France. https://www.hdigitag.fr/wp-content/uploads/Laurens-Klerkx-SeminaireDigitAgSept252019-Social-Science-on-digitalization-in-Ag.pdf

Klerkx L (2020) Advisory services and transformation, plurality and disruption of agriculture and food systems: towards a new research agenda for agricultural education and extension studies. J Agric Educ Ext 26(2):131–140. https://doi.org/10.1080/1389224X.2020.1738046

Lave J, Wenger E (1990) Situated learning: legitimate peripheral participation. Cambridge University Press, Cambridge

Lowder SK, Skoet J, Raney T (2016) The number, size, and distribution of farms, smallholder farms, and family farms worldwide. World Dev 87:16–29. https://doi.org/10.1016/j.worlddev.2015.10.041

Matos F, Vairinhos V, Salavisa I, Edvinsson L, Massaro M (2020) Introduction. In: Matos F, Vairinhos V, Salavisa I, Edvinsson L, Massaro M (eds) Knowledge, people, and digital transformation: approaches for a sustainable future. Springer, Cham, pp 1–6

Matthews JR (2017) Understanding indigenous innovation in rural West Africa: challenges to diffusion of innovations theory and current social innovation practice. J Hum Dev Capabil 18(2):223–238. https://doi.org/10.1080/19452829.2016.1270917

Mohapatra S (2020) Gender differentiated economic responses to crises in developing countries: insights for COVID-19 recovery policies. Rev Econ Househ. https://doi.org/10.1007/s11150-020-09512-z

Morgan SL (2011) Social learning among organic farmers and the application of the communities of practice framework. J Agric Educ Ext 17(1):99–112. https://doi.org/10.1080/1389224X.2011.536362

Narine L, Harder A (2019) Extension officer's adoption of modern information communication technologies to interact with farmers of Trinidad. J Int Agric Ext Educ 26(1):17–34. https://doi.org/10.5191/jiaee.2019.26103

Norton GW, Alwang J (2020) Changes in agricultural extension and implications for farmer adoption of new practices. Appl Econ Perspect Policy 42(1):8–20. https://doi.org/10.1002/aepp.13008

Nuutinen M, Filho WL (2018) Online communities of practice empowering members to realize climate-smart agriculture in developing countries. In: Azeiteiro UM, Leal Filho W, Aires L (eds) Climate literacy and innovations in climate change education: distance learning for sustainable development. Springer, Cham, pp 67–83

Oreglia E (2014) ICT and (Personal) development in rural China. Inf Technol Int Dev 10(3):19–30

Orlikowski W (2000) Using technology and constituting structures: a practice lens for studying technology in organizations. Organ Sci 11:404–428. https://doi.org/10.1287/orsc.11.4.404.14600

Pereira L, Wynberg R, Reis Y (2018) Agroecology: the future of sustainable farming? Environ Sci Policy Sustain Dev 60(4):4–17. https://doi.org/10.1080/00139157.2018.1472507

Phillips PWB, Relf-Eckstein J-A, Jobe G, Wixted B (2019) Configuring the new digital landscape in western Canadian agriculture. NJAS Wagening J Life Sci 90–91:100295. https://doi.org/10.1016/j.njas.2019.04.001

Remolina N, Findlay MJ (2021) The paths to digital self-determination – a foundational theoretical framework. SMU Centre for AI & Data Governance Research Paper No. 03/2021. https://ssrn.com/abstract=3831726

Rotz S, Gravely E, Mosby I, Duncan E, Finnis E, Horgan M et al (2019) Automated pastures and the digital divide: how agricultural technologies are shaping labour and rural communities. J Rural Stud 68:112–122. https://doi.org/10.1016/j.jrurstud.2019.01.023

Shilomboleni H, Pelletier B, Gebru B (2020) ICT4Scale in smallholder agriculture: contributions and challenges. Inf Technol Int Dev 16:47–65

Steinke J, van Etten J, Müller A, Ortiz-Crespo B, van de Gevel J, Silvestri S, Priebe J (2020) Tapping the full potential of the digital revolution for agricultural extension: an emerging innovation agenda. Int J Agric Sustain:1–17. https://doi.org/10.1080/14735903.2020.1738754

Stordy P (2015) Taxonomy of literacies. J Doc 71(3):456–476. https://doi.org/10.1108/JD-10-2013-0128

Tacchi J (2015) Ethnographic action research: media, information and communicative ecologies for development initiatives. In: Bradbury H (ed) The SAGE handbook of action research, 3rd edn, pp 220–229. SAGE Publications, London https://doi.org/10.4135/9781473921290

Tran TA, James H, Pittock J (2018) Social learning through rural communities of practice: empirical evidence from farming households in the Vietnamese Mekong Delta. Learn Cult Soc Interact 16:31–44. https://doi.org/10.1016/j.lcsi.2017.11.002

Triste L, Debruyne L, Vandenabeele J, Marchand F, Lauwers L (2018) Communities of practice for knowledge co-creation on sustainable dairy farming: features for value creation for farmers. Sustain Sci 13(5):1427–1442. https://doi.org/10.1007/s11625-018-0554-5

Vyas D, Chisalita CM, Veer GCvd (2006) Affordance in interaction. Paper presented at the proceedings of the 13th Eurpoean conference on cognitive ergonomics: trust and control in complex socio-technical systems, Zurich. https://doi-org.login.ezproxy.library.ualberta.ca/10.1145/1274892.1274907

Vyas D, Chisalita CM, Dix A (2016) Organizational affordances: a structuration theory approach to affordances. Interact Comput 29(2):117–131. https://doi.org/10.1093/iwc/iww008

Walsham G (2020) South-South and triangular cooperation in ICT4D. Electronic Journal of Information Systems in Developing Countries 86(4). https://doi.org/10.1002/isd2.12130

Wanigasundera WADP, Atapattu N (2019) Extension reforms in Sri Lanka: lessons and policy options. In: Babu SC, Joshi PK (eds) Agricultural extension reforms in South Asia. Academic Press/Elsevier, London, pp 79–98

Wenger E, White N, Smith JD (2009) Digital habitats: stewarding technology for communities. CPSquare, Portland

Chapter 10
Connectivity Literacy for Digital Inclusion in Rural Australia

Amber Marshall, Rachel Hay, Allan Dale, Hurriyet Babacan, and Michael Dezuanni

Introduction

In Australia, 28% of the population (around 7 million people) live outside major cities (AIHW 2020). Rural areas are predominantly occupied by farmers and Indigenous traditional owners. Farmers are stewards of some 51% of Australia's land area (Australian Bureau of Statistics [ABS] 2018a)[1]; other key industries include Indigeneous enterprise, forestry, tourism, mining, health, education, and other government services. In the State of Queensland, in which this study took place, 85.9% of the land is for cattle grazing (Queensland Government 2018), with some properties exceeding one million acres. Farming households and communities in rural Queensland are thus often located many hundreds of kilometres from cities and regional centres.

Owing to this 'tyranny of distance' rural residents often do not have access to the high-quality, affordable, and reliable telecommunications services that are more readily available in urban settings. Research also shows that rural Australians tend

[1] For the purposes of this study, 'rural' includes three levels of the Australian Bureau of Statistics' (ABS) remoteness structure (ABS 2018b): Outer Regional, Remote Australia and Very Remote Australia. References to 'urban' areas and populations include the remaining two levels of the ABS' remoteness structure: Inner Regional Australia and Major Cities of Australia.

A. Marshall (✉)
Griffith University, Nathan, QLD, Australia
e-mail: amber.marshall@griffith.edu.au

R. Hay · A. Dale · H. Babacan
James Cook University, Townsville/Cairns, QLD, Australia

M. Dezuanni
Queensland University of Technology, Brisbane, QLD, Australia

D. Radovanović (ed.), *Digital Literacy and Inclusion*,
https://doi.org/10.1007/978-3-031-30808-6_10

to have fewer digital skills to enable them to use digital connections and technologies to their full advantage (Marshall et al. 2020; Park 2017). This is a problem because a large portion of civic, social, and business-related activities must now be conducted online, including on grazing and other farming properties.[2] For example, digital exclusion can impact rural partnerships and undermine development (Erdiaw-Kwasie and Khorshed 2016) and inhibit access to education, emergency communication, and health services for rural families (Freeman et al. 2016).

In 2016, the United Nations declared access to telecommunications (including mobile and internet) a human right (Szoszkiewicz 2018). Not only must access to the internet be available; it must also be affordable. Moreover, digital skills are increasingly being considered as essential to overall literacy in the modern world (Livingstone et al. 2021; Radovanovic et al. 2020; Mills 2010). Yet, according to the Australian Digital Inclusion Index [ADII] (Thomas et al. 2021), 'regional' Australians score significantly lower on all three sub-indices (access, affordability, digital ability) than 'metro' Australians. Moreover, historically 'farmers and farm managers' (as defined by the ABS and ADII) have been particularly digitally disadvantaged for reasons that are still not fully understood (Marshall et al. 2020), though progress is being made. The recent COVID-19 pandemic has also highlighted and exacerbated the inherent social and economic disadvantage of people who lack the means to reliably access and use digital technologies (Thomas et al. 2020), such as those in rural Australia.

This chapter discusses a project that investigated the underlying factors of low levels of digital inclusion in rural households and communities in the Gulf Savannah region of Far North Queensland (FNQ). We used qualitative methods (ethnographic interviews, focus groups, and participant observation) to give a voice to farmer participants. While we set out to investigate participants' challenges associated with access, affordability, and digital ability, *connectivity literacy* was identified as an extra layer of digital knowledge and skills required of rural consumers to navigate the complex telecommunications environment. Connectivity literacy was first defined by the Better Interent for Rural, Regional and Remote Australia (BIRRR) advocacy group as *'all the skills and knowledge needed by a consumer to get connected and stay connected to telecommunications services'*) (Hartsuyker et al. 2021 p.9). It is distinct from digital literacy, which pertains to knowledge and skills required to navigate digital devices and the internet effectively and confidently (Lankshear and Knobel 2008) once the internet connection has been made. Connectivity literacy is about getting connected to digital devices and broadband connections in the first place.

We begin the chapter by briefly reviewing relevant academic and grey literature in relation to digital inclusion in rural Australia, before providing a contextual background and methodology for the study. We then identify the factors of rural digital inclusion as they emerged around the themes of access, affordability and digital

---

[2] Other significant rural populations include Indigenous and mining communities, but these were not the focus of this particular study.

ability, and articulate several aspects of connectivity literacy that are essential for rural digital inclusion. We conclude by describing how connectivity literacy adds a new, cross-cutting dimension to digital inclusion as it may be understood and applied in rural contexts.

Digital Inclusion in Australia

Digital inclusion is related to the more familiar term 'digital divide', which has traditionally referred to the presence or absence of telecommunications and internet infrastructure and access (Compaine 2001). Recently, however, researchers have begun to adopt the more holistic term 'digital inclusion', which recognises the varying degrees to which people might access the internet, as well as how affordable and reliable the service is, and people's capacity to put the internet to work in everyday life (Ragnedda and Mutsvairo 2018). Links have been established between digital inclusion and other forms of inclusion. Socially disadvantaged groups (e.g., older people, low-income earners, and people with a disability) are less likely to have affordable access to the internet or have the necessary skills to make effective use of connections (Helsper 2008). The reverse is also true; digital disadvantage can exacerbate socioeconomic disadvantage. For example, not being connected can lead to difficulty applying for jobs, using government services, and accessing health information online, which can degrade one's personal circumstances.

A key measurement of digital inclusion in Australia is the ADII (Thomas et al. 2021). An annual nationwide survey is undertaken across three indices: access (comprised of interest access, use of internet technology, and data allowance); affordability (comprised of relative expenditure of household income on access, and value of expenditure on data allowance); and digital ability (comprised of attitudes to technology, basic skills, and online activities). The Index is predicated on the notion that the 'perfectly included' individual would score 100. In 2018 (the year data were collected for this paper), the average national score across the sub-indices was 60.2 (Thomas et al. 2018) and in 2019 the national average was 61.9 (Thomas et al. 2019). Results for specific populations reflect the link between digital inclusion and social inclusion. In 2018 people aged over 65, for example, scored 46.0, and those not in the labour force scored 52.0. Of relevance to our study is that in 2018 people in rural areas scored 53.9, which is 8.5 points below those in capital cities.[3] Also, in 2017/2018 the average score for 'farmers and farm managers' was 45.4 compared with the rural average of 52.8 and national average of 59.2 (Marshall et al. 2020).

Several researchers have explored geographically based digital inclusion in Australia. Willis and Tranter (2006) observed that lack of basic telecommunications

[3] In 2021, the national average had risen to 71.1, with metro areas scoring 72.9 and regional areas scoring 67.4 (−5.5). Though the urban/rural divide is narrowing, it has persisted particularly in the dimension of digital ability (Thomas et al. 2021).

and internet infrastructure largely underpined low levels of digital inclusion in rural areas. Park (2017: 399) found that 'remoteness was a strong indicator of digital exclusion' and that rural populations generally have lower levels of education and employment that can impede digital participation (Park 2017). Freeman et al. (2016) identified that business development, education, emergency communication, and health were key drivers for broadband internet connections for socioeconomic development in rural Australia. Overall, these authors demonstrate the broad range of inter-connected technical, social, and economic factors impacting digital inclusion in rural areas, which are also reflected in international contexts (Salemink et al. 2017).

Telecommunications in Rural Australia

Understanding the development of telecommunications in Australia provides context to macro-level factors leading to poor digital inclusion in rural areas. Telecommunication infrastructure and services (landlines, broadband, mobile) transitioned from public to private providers in Australia from the 1990s to the 2000s. However, Australia's broadband internet needs in the emerging digital economy were not adequately met (Park 2017). In response, the Australian Government's National Broadband Network (NBN) aimed to bring high-speed broadband into every Australian household (Rudd et al. 2007). Following several government and policy changes, the final NBN solutions for rural areas were fixed wireless and satellite (NBN Co Ltd 2019), which are generally slower, less reliable, provide less data, and offer less value for money than fixed line NBN connections that are mostly in urban areas (Australian Government 2018a). These and other issues, such as high latency, data caps, and poor customer service, are well-documented problems for rural NBN users (Hay 2018). Meanwhile, mobile phone and mobile broadband services in rural Australia have been characterised by blackspots, outages, and oversubscription (Marshall et al. 2020). While the country's major mobile carrier purports to achieve over 99% 3/4G coverage of the Australian population (Simpson, 2017), this is restricted to about one third of the continent's landmass.[4]

Tri-annual Regional Telecommunications Reviews are contributing to improved digital connectivity in rural Australia. Outcomes of three recent reviews (Hartsuyker et al., 2021; Edwards et al. 2018; Shiff et al. 2015) have resulted in: increased investment, by governments and telecommunications providers, into regional digital connectivity (e.g., Regional Connectivity Program (RCP) (Australian Government n.d.b) and Mobile Blackspot Program (MBSP) (Australian Government n.d.a)); transparency around the NBN rollout and its performance; increased safeguards for the modern telecommunications environment; and a Regional Tech Hub (free online/phone help desk) to assist with rural digital inclusion (McKenzie 2018). The Regional Connectivity Program (Australian Government 2021) invests in the delivery of place-based telecommunications infrastructure projects to improve digital

[4] https://www.telstra.com.au/coverage-networks/our-coverage.

connectivity across regional, rural, and remote Australia. Funded projects include alternative fixed wireless broadband operations, mobile voice and data solutions, upgrades to fibre backhaul, deployment of fibre, and satellite broadband. While the choice of products and providers has increased significantly in the recent years (leading to increased speeds, higher data limits, addition of unmetered data products, and month-by-month contracts), additional connectivity options have made it harder for consumers to navigate solutions to meet their needs. Therefore, a lack of connectivity literacy will still inhibit consumers from getting online.

It is important to acknowledge that, despite great challenges, many rural Australians have been able to access and use digital connections and technologies in business and everyday life, including in farming communities. Hay and Pearce (2014) show, for example, that rural women are driving the adoption of digital agriculture technologies (e.g., remote cameras, remote weather stations, livestock monitoring) in Northern Australia by transitioning from computer-based tasks in the home/office to implementing digital connectivity and technology in the paddock. Moreover, Marshall (2021) shows how 'farm wives' are further translating their homestead-related digital skills into other activities, such as online businesses and social media collaboration. These studies reflect the great tenacity and ingenuity of rural Australian women (Alston 1995, 1998), which has extended from adoption of early telecommunications and machinery to mobile and broadband technologies of today. Our study of digital inclusion in rural Australia, and the articulation of connectivity literacy, somewhat reflects this culture of coping and problem-solving among rural women and men.

Contextual Background and Methodology

Our study was conducted in rural towns and on grazing properties in Australia's Gulf Savannah region. We partnered with Gulf Savannah Natural Resource Management (formerly Northern Gulf Resource Management Group) to undertake this research. As indicated (approximately) by the green area in Fig. 10.1, the area extended from Mount. Molloy to Normanton (about 600 km) and included the Local Government Areas of Mareeba Shire (part), Croydon Shire, Etheridge Shire, Cook Shire (part), Carpentaria Shire (part) and Kowanyama Shire. Ninety per cent of this area is grazing land dominated by about 160 large pastoral leases (Gulf Savannah NRM n.d.). In 2018, the ADII score for the area was 52.8 (Fig. 10.2), 6.1 points below the state of Queensland and 9.6 points below Australia's capital cities.

Fieldwork was undertaken just from June to October 2018,[5] before the COVID-19 pandemic, enabling researchers to attend rural events and make face-to-face home

[5] Even though the data was collected in 2018 our ongoing work in the region (Marshall et al. 2023) means we can confirm that the insights are still, if not more, relevant now owing to the ways in which the pandemic has exacerbated the digital divide, and the increasing digitalization of social and commercial services.

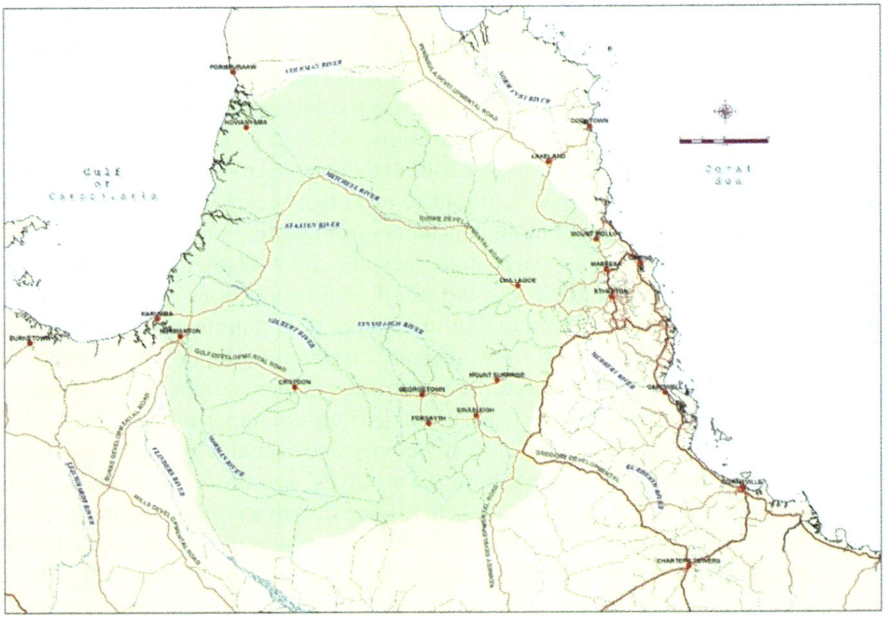

Fig. 10.1 The Savannah Gulf region (https://plan.northerngulf.com.au/our-region/)

visits to remote farming homesteads. Data collection took place over three separate week-long field trips where a researcher worked alongside the local industry partner who provided introductions to participants and opportunities to attend events. Ethnographic interviews, participant observation, and focus groups (Flick 2006) were employed to document the participants' interactions with digital technologies and each other. Flick (2006: 166) notes that while a specific time and place is usually arranged for 'regular' interviews, an ethnographic approach gives rise to spontaneous opportunities for participant engagement arising from regular field contact. Accordingly, interviewees were selected using convenience sampling, and interviews were conducted in locations such as town halls, paddocks, and home kitchens. Data collection activities included: ten audio-recorded in-person interviews (one-to-two people per interview); two audio-recorded focus groups; participant observation at four rural events (women's workshop, digital platforms for agriculture workshop, and dung beetles for soil health workshop; bush business seminar); and ad hoc interviews with several other participants. Participants (n => 30) were a combination of men and women ranging in age from their 20s to 60s. Given the nature and timing of the events, the majority of participants were women. Nvivo (version 12) was used to undertake thematic coding of the data (Flick 2006) according to the ADII's three dimensions of digital inclusion: access, affordability, and digital ability.

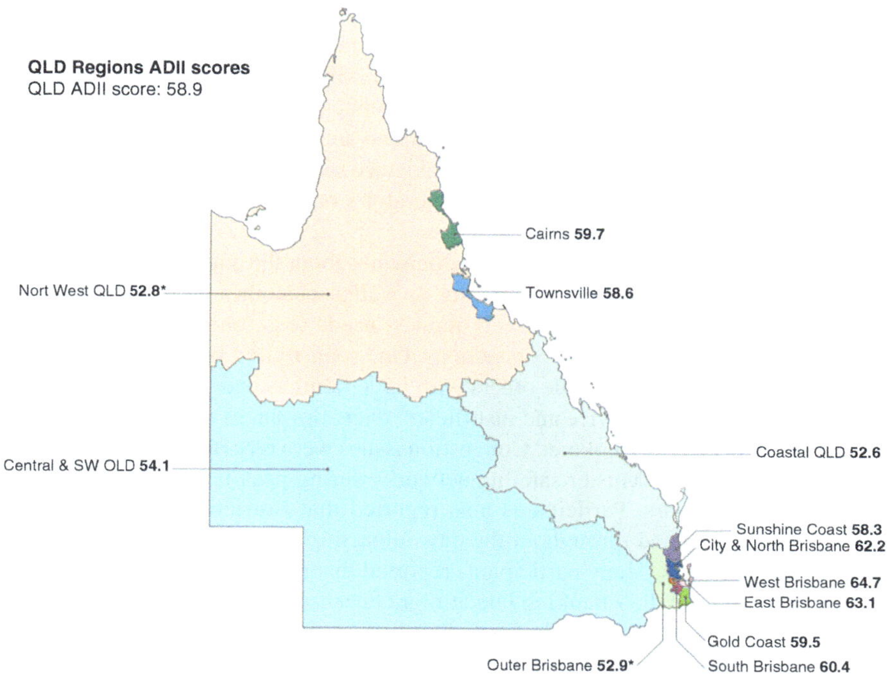

Fig. 10.2 Australian Digital Inclusion Index (ADII) scores in Queensland in the year data was collected. (Thomas et al. 2018: 30)

We present the results of the study[6] under each of the three dimensions of digital inclusion by identifying the enabling/constraining factors of digital inclusion through the lived experience of participants; and making observations about the kinds of specific skills and knowledge that emerge in rural consumers' efforts to affordably and reliably get and stay connected, and to use these connections meaningfully in rural life.

Accessing Digital Connections

Gulf Savannah residents were connected to the internet and mobile services in different ways depending on their location. People within approximately 5–10 km of a town generally had 3G or 4G mobile coverage and/or access to NBN fixed wireless (ADSL connections also existed where NBN broadband was not yet available).

[6] All personal and property names have been anonymised.

More remote residents connected to the internet through the NBN Sky Muster satellite, or by mobile networks when in town. Many carried a satellite phone (or smart phone with a satellite sleeve) when travelling or working in no-service areas, such as when mustering cattle. Access was also dependent on remote power infrastructure, which is typically less reliable than in urban areas and exposes consumers to a greater risk of internet outages. As one interviewee said, ‘It (the internet connection) is satellite, and it only works when the generator’s on anyway, because you need a power source’.

There were divergent views among participants about the adequacy of these connections in terms of reliability and quality, as well as customer service and installation. Some participants with relatively modest needs (e.g., one couple household) had few complaints about their connections. One semi-retired farming women said, ‘basically, we can get by’, while others with large families and/or work crews complained that unreliable service and insufficient data allowances meant their business and personal needs were not met. Congestion issues were reported as being acute on mobile and NBN Sky Muster satellite networks during peak times, especially during after-school hours. Participants also reported that restrictions to internet and mobile access changed throughout the day, impacting how and when people completed their activities. Many participants reported that they responded to customers, paid wages, and did their banking late at night because web-based platforms could not be reliably accessed during business hours. Recreational internet activities were also limited. For example, people said they could not watch videos or streaming services in the usual ‘on demand’ format. As one participant said, ‘I download my show when I get up at six in the morning so I can watch it later on after dinner’.

Participants reported that trouble-shooting technical faults caused immense stress and frustration. Because of their remote location, telecommunications providers often cannot readily deploy technicians (or there are lengthy delays), which means customers must try to fix issues on their own over the phone with a provider representative. Many participants acknowledged that services had improved since the launch of NBN Sky Muster and the federal government’s Mobile Black Spot Program. However, the time and resources required to tend to regular connection issues (including lost productivity) were a major concern impacting all aspects of life.

This state of affairs has, in many cases, spurred rural consumers to adopt new knowledge, skills, and practices to accommodate unreliable/insufficient connections, which may be understood as connectivity literacies. Such literacies include: knowing which routers/modems work best with satellite/fixed wireless connections; configuring local hardware (e.g., modem, router) for optimal connection to broadband service (e.g., fixed wireless, satellite); installing extra hardware (e.g., boosters) to make connections work optimally; purchasing and configuring ‘unconventional’ devices (e.g., satellite phones/sleeves, Telstra blue tick smart phones); and fixing issues related to critical power connections (e.g., generator).

Affordability of Digital Connectivity

Given the above-described sub-optimal telecommunications services, we found that affordability and value for money for consumers were highly impacted. Participants reported that when providers assured them that services were 'live' in their area, the reality was often quite different, owing to context-specific factors such as hills in the line of sight of the nearby tower (for mobile or fixed wireless connections). Participants also reported extended outages and waiting weeks and even months for maintenance and support. One interviewee said, 'I don't (report outages). I can't be bothered. It's a waste of time because they're not going to do anything about it'. Rural residents often continued to pay for services even when they were interrupted; they would not risk stopping their payments in case this incurred further interruptions. While providers do offer credit or refunds in some instances, the frustration expressed by some participants in dealing with providers and attempting to resolve issues was palpable. Call centre workers based in urban areas often do not understand the contextual factors that impact rural services, leading to unhelpful and incorrect advice being provided.

Gulf Savannah residents also paid more for less in terms of telecommunications hardware and service. While there is national pricing for mobile plans (all end-users, regardless of location, are charged the same), value for money is unequal across rural and urban populations. This is because fewer towers are available in rural areas to provide services, and consistency of coverage is superior in urban areas. To compensate for the lack of networked connectivity, many participants purchased a combination of services (3G/4G, satellite, fixed wireless) and devices (mobile phone, desktop computer, tablet) to create redundancy. For example, some people said they had two mobile phones (with a different sim card in each, e.g., Telstra and Optus)[7] so that they could connect to the local network wherever they were travelling within the region. Others relied on two-way radios or carried satellite phones or sleeves for areas with no mobile coverage at all, which participants said were often expensive and unreliable. Moreover, participants had limited choice of service providers operating in sparsely populated areas. Consequently, participants 'layered-up' on several services and devices, in the hope one would work at any given time, which greatly compounded the cost of being connected (Marshall et al. 2019).

Broader consumer issues that impacted Gulf Savannah residents were also identified. Namely, there seemed to be a mismatch between consumer needs and telecommunications services in the region. For example, terms and conditions of internet plans (e.g., on/off-peak data allocations) did not lend themselves to routine technology uses in participants' households and businesses. In addition, mobile phone plan structures did not accommodate heavy use in short increments, which was common among the participants. For example, contract cattle mustering teams live and work in camps often located at considerable distance from the homestead

[7] Dual sim cards were either unavailable or not known to the participants.

(one person said 60 km), with non-existent or very limited mobile and broadband connection. One interviewee said she only used her mobile phone regularly for 4 months of the year, but an expensive, unlimited plan was needed so she could do everything she wanted to when in range. Information and support for choosing a suitable provider and product was also difficult to attain for participants, who often relied on recommendations from friends, or simply stuck with their default provider and product for fear of causing more issues if they switched.

Several connectivity literacies emerged as being critical to rural consumers' being able to attain economical mobile and broadband services. For example, people needed to be able to find information about plans suited to rural needs. Providers don't tend to advertise rural products, as there are far more urban consumers to sell to. In Australia this has led to third parties (e.g., BIRRR, Isolated Children's Parents Association, National Rural Health Alliance, and other volunteer groups) curating information for rural consumers. Other connectivity literacies required by rural consumers to help ensure they get an affordable and fair connection/service include: actively monitoring and managing speed/data use across several users and devices; configuring devices to update during off-peak periods to get value for money and avoid service interruption; and knowing their telecommunications consumer rights and advocating for them with providers and policymakers.

Using Digital Technologies

Digital ability, somewhat synonymous with digital literacy but differing from connectivity literacy, refers to the knowledge and skills required to use devices (computer, laptop, smart phone, tablet, IoT sensor) and software (operating systems, apps, web browsers, cloud-based platforms) to access the internet and web-based information and services. One interviewee recounted the various digital tasks associated with day-to-day business on the farm:

> It's just, all our programs now are on the computer. You know, it's just, it's how you do your banking, all your programs come through on it. It's just how you live. All your invoicing comes through on it.... All your cattle programs, and even my bookwork that I do; it's all internet and then it gets sent down to head office.

Important information for farmers—such as market trends, policies, regulations, and processes—are all changing rapidly and are mostly online. Pastoralists must complete online training modules to attain and maintain their Livestock Production Assurance accreditation, which involves accessing video, audio, images, and textual content. Some participants found it difficult, and sometimes impossible, to complete these modules owing to poor reliability of access, slow download speeds, and data restrictions. Other participants reported that they found it difficult to access property and vegetation maps to comply with vegetation clearing laws, which are most readily available online. Others were resisting using the online National Livestock Identification System used to track livestock, owing to a lack of interest in shifting from the familiar paper-based system. Overall, the research showed that some participants lacked rudimentary ability to access and use devices and the internet for these essential tasks.

The research further provided some nuanced insights as to why some people were willing and able to adopt digital technologies and others were not. First, gender seemed to play an important role. Women who often 'ran the house' were responsible for managing scarce household resources, which included internet access and data consumption. This was particularly challenging in households that relied on NBN Sky Muster satellite plans,[8] which were capped and shaped, or those using mobile-only data. Participant women reported that the total maximum monthly download was stretched thinly across several people (adults, children, workers, visitors) and priorities (social life, business, schooling). Moreover, data was often used quickly by visitors or workers who were unaware of (or did not respect) the property's internet limitations. One interviewee described how her adult children expend satellite internet data when they come home to work or visit:

> (My husband) is on the internet checking for emails, then he's looking for a dozer (bulldozer) for sale. Then the boys (adult children) are on YouTube ... So then when data stops, (he) gets there and he's saying "Right, it was only renewed a week ago and I can't get emails. I can't this, I can't get that." Everything is so slow.

In the face of this scarcity, some women developed ways to allocate and monitor data use. One mother, for example, said she allocated specific amounts of data to the farm business, children's education, and parents' social life and then allowed her children to 'go nuts' on YouTube on the seventh day of every month before the new cycle began (if any data remained). Another more digitally savvy interviewee said she and her husband used an app to track data usage by all devices on the WiFi network; they also changed their password occasionally to prevent visitors logging on automatically.

Second, age of family members was a factor in understanding digital ability in farming households, where it is common for multiple generations to co-habit in one or several dwellings on a single property. Some participants in the middle generation (30s and 40s), who were educated in regional or city-based high schools and/or universities, had acquired digital knowledge and skills and an appreciation for the uses of technology on properties. Their parents (in their 60s and 70s), however, often did not share the same understanding or enthusiasm for digital technologies. For example, on one property in the study, the parents/owners had agreed to implementing livestock RFID (radio frequency identification) for cattle tracking, but the younger property managers did not have the authority to spend money to make further investments to enable RFID data to be shared between devices and the cloud. While age is a known factor in determining attitudes towards digital technologies, this seemed to be pronounced among the participant group.

Digital ability among the research participants was closely related to several features of connectivity literacy we have already identified. For example, under 'accessibility' we noted the need to actively monitor and manage speed/data use across

[8] NBN Sky Muster plans have improved in terms of data allowance and metering since data was collected for this research.

several devices. Here, by extension, connectivity literacy can also involve developing elaborate, frugal, and creative ways to allocate and monitor data use according to completing needs and demands of the business, parents, children, workers, and visitors (e.g., manual or digital roster system for access and data). As well, in the absence of local technical support in remote areas, rural people (in this case, particularly younger people and women) often needed to mentor/educate others in their family or community. Such digital mentoring is a difficult and time-consuming task requiring a suite of skills to be done effectively (Dezuanni et al. 2019).

Table 10.1 shows a summary of the findings regarding the contextualised factors that contribute to low levels of digital inclusion among Gulf Savannah residents and the aspects of connectivity literacy abstracted from the findings.

Table 10.1 Summary of factors and impacts of low levels of digital inclusion in Gulf Savannah pastoral families and communities

Dimension of digital inclusion	Factors of digital inclusion	Aspects of connectivity literacy
Access	People still rely on traditional telecommunications, such as telephone landlines and two-way radios Speed, congestion, and unreliability of connections are issues for some Remote critical energy infrastructure makes rural areas vulnerable to outages Access issues relate particularly to the network, not just nodes	Knowing which routers/modems work best with satellite/fixed wireless connections Configuring local hardware (e.g., modem, router) for optimal connection to broadband service (e.g., fixed wireless, satellite) Installing extra hardware (e.g., booster) to make connections work optimally Purchasing and configuring 'unconventional' devices (e.g., satellite phones/sleeves, Telstra blue tick smart phones) Fixing issues related to critical power connections/generator. People keep unconventional hours to get work done online
Affordability	Telecommunications coverage is often patchy and unreliable, particularly on the periphery of towns There are less services and less coverage in rural areas Terms and conditions of services do not lend themselves to rural modes of use Rural residents 'layer up' on hardware and services, which is expensive	Finding information about plans suited to rural needs Actively monitoring and managing speed/data use across several devices Configuring devices to update during off-peak periods to get value for money and avoid service interruption Knowing telecommunications consumer rights and advocating for them with providers and policymakers
Digital ability	Data is often a contested resource managed by women Many men seemed to lack interest and skills in digital technologies There are inter-generational digital literacy challenges	Developing elaborate, frugal, and creative ways to allocate and monitor data use Mentoring others in the use of digital technologies

Since this research project was undertaken, there has been greater investment in mobile and broadband infrastructure and consequent incremental improvements to mobile services and upgrades to fibre broadband connections in rural Australia. Notably, NBN announced expansion of its fixed wireless network, which draws customers away from satellite services on to superior fixed wireless services, which in turn relieves pressure on the satellite service for remaining customers. There has also been consistent bi-partisan investment in mobile and broadband infrastructure through the MBSP and RCP. Looking forward, in the 2023-24 federal budget, the Australian Government committed $AU656 million to its Better Connectivity Plan for Regional and Rural Australia, which includes: boosting multi-carrier mobile coverage on regional roads, investing in on-farming connectivity programs, an independent audit of mobile coverage to better identify black spots and guide investment priorities; and further support for existing infrastructure and capacity-building programs like the MBSP, RCP and RTH (King 2022). Nonetheless, the parts of the agricultural sector continues to report challenges to attaining the necessary 'connectivity threshold' to enable and sustain farms into the future (Australian Broadband Advisory Council 2021).

Conclusion

This chapter provides insights into the complexities of getting connected and staying connected to telecommunications services in rural Australia. It calls on a case study from the Gulf Savannah region of Far North Queensland to identify and understand how connectivity literacy impacts low levels of digital inclusion. Key measures of digital inclusion (access, affordability, and digital ability), while helpful in benchmarking digital and social inclusion across populations, do not adequately acknowledge or measure the skills required by individuals to achieve reliable, affordable connectivity in rural areas. We have described, and evidenced with primary data, *connectivity literacy* (in rural areas in particular) as being principally about being able to: set up local hardware and networks; respond to technical outages when they occur; and navigate a complex consumer environment to ascertain the best connectivity options. Knowledge about accessibility, affordability, and digital ability (online skills) is unlikely to assist in these critical tasks.

With telecommunications providers increasingly centralising their support services, many rural residents are located long distances away from help, which often means they must attempt to fix connectivity issues themselves. This added pressure, on top of usual daily duties, creates high stress and costs in rural users in planning, deciding, connecting, up/downgrading, and changing telecommunications provers and products. Those with limited connectivity literacy first seek assistance from their providers whose call centre operators often have limited rural connectivity literacy themselves, and are not always transparent in their solutions. As a result, rural users seek assistance from others in their local network; then those with high connectivity literacy may come to (be expected to) provide technical support to other families in community, causing further interruptions and stress in their own life and business. A critical component of this connectivity literacy is keeping up with product offerings (e.g., new phones, boosters, plans) and their associated costs. Moreover, understanding the intricacies of how data can be allocated within service plans, and which products can be layered up (or not), takes considerable time and tenacity. Connectivity literacy is, therefore, a substantial skill and, without it, navigation of the digital connectivity environment is frequently fraught with frustration and disappointment.

As the need to conduct civic, social, educational, and business-related activities online increases, the demand for connectivity literacy increases. While the pressure to get connected in an increasingly complex rural technology environment can overwhelm some rural consumers, many (in this and other studies) display considerable tenacity to problem-solve access to and use of digital connections (for example, transitioning to digital farming, smart phones, plug and play devices, different versions of software, cloud-based storage, and the Internet of Things). While the Australian Government and telecommunications companies are making significant investment towards connectivity infrastructure and place-based solutions, there is little investment in human capacity around connectivity literacy. Targeted programmes should be created to foster connectivity literacy development within and between rural households and communities. Research (Dezuanni et al. 2019) suggests that adult learners benefit from the social connections that come with community-led, peer-to-peer support. Accordingly, ‘digital ranger’ programmes could employ local, digitally savvy people to provide group training and/or one-on-one assistance to help others get connected and troubleshoot issues. Like agricultural R&D extension officers, digital rangers could be employed in local communities or deployed from regional centres through government or industry-funded schemes. Increased digital connectivity and capability could empower individuals to improve their social and economic circumstances, enhance liveability in rural communities, and bolster regional development in Australia.

Acknowledgements This study was a partnership between James Cook University and Gulf Savannah Natural Resource Management Group (formerly Northern Gulf Resource Management Group), funded by the Australian Communications Consumer Action Network (ACCAN).

References

Alston M (1995) Women on the land: the hidden heart of rural Australia. UNWS Press, Kensington

Alston M (1998) Farm women and their work: why is it not recognised? J Sociol 34(1):23–34

Australian Broadband Advisory Council (2021) Agri-Tech Expert Working Group: report. Australian Government. https://www.infrastructure.gov.au/department/media/publications/agri-tech-expert-working-group-report

Australian Bureau of Statistics (ABS) (2018a) Land management and farming in Australia, 2016–17, cat No. 4627.0. ABS, Canberra

Australian Bureau of Statistics (ABS) (2018b) Remoteness structure. ABS, Canberra. https://www.abs.gov.au/websitedbs/d3310114.nsf/home/remoteness+structure

Australian Government (n.d.a) Mobile black spot program. Department of Infrastructure, Transport, Regional Development and Communications, Canberra. https://www.communications.gov.au/what-we-do/phone/mobile-services-and-coverage/mobile-black-spot-program
Australian Government (n.d.b) Regional connectivity program. Online: Australian Government. https://www.infrastructure.gov.au/media-technology-communications/internet/regional-connectivity-program
Australian Institute of Health and Welfare (2020) Rural and remote health. https://www.aihw.gov.au/reports/australias-health/rural-and-remote-health
Compaine BM (ed) (2001) The digital divide: facing a crisis of creating a myth? MIT Press, Cambridge, MA
Dezuanni M, Marshall A, Cross A, Burgess J, Mitchell P (2019) Digital mentoring in Australian communities—a report prepared for Australia Post. Australia Post, Australia
Edwards S, Duncan W, Plante J, Sefton R, Stretton K, Weller P (2018) 2018 regional telecommunications review: getting it right out there. Australian Government, Canberra. https://www.infrastructure.gov.au/media-centre/publications/2018-regional-telecommunications-review-getting-it-right-out-there
Erdiaw-Kwasie MO, Khorshed A (2016) Towards understanding digital divide in rural partnerships and development: a framework and evidence from rural Australia. J Rural Stud 43:214–224
Flick U (2006) An introduction to qualitative research, 3rd edn. Sage, London
Freeman J, Park S, Middleton C, Allen M (2016) The importance of broadband for socio-economic development: a perspective from rural Australia. Australas J Inf Syst 20:1–18
Gulf Savannah Natural Resource Management Group (n.d.) Our region. https://gulfsavannahnrm.org/our-region/
Hartsuyker L, Sparrow K, Middleton S, Bradlow H, Cosgrave M (2021) 2021 regional telecommunications review—a step change in demand. Australian Government, Canberra. https://www.infrastructure.gov.au/department/media/publications/2021-regional-telecommunications-review-step-change-demand
Hay R (2018) Better internet for rural regional and remote Australia: landline & connectivity survey results, 2018 (ISBN: 978-0-9954471-8-9). https://birrraus.com/category/survey/
Hay R, Pearce P (2014) Technology adoption by rural women in Queensland, Australia: women driving technology from the homestead for the paddock. J Rural Stud 36:318–327
Helsper E (2008) Digital inclusion: an analysis of social disadvantage and the information society. Department for Communities and Local Government, London
King E (2022) Regional telecommunications measures: budget review 2022–23 index, April 22. Parliament of Australia. https://www.infrastructure.gov.au/media-communications-arts/better-connectivity-plan-regional-and-rural-australia
Lankshear C, Knobel M (2008) Introduction. In: Lankshear C, Knobel M (eds) Digital literacies: concepts, policies and practices. Peter Lang, New York, pp 1–16
Livingstone S, Mascheroni G, Mascheroni M (2021) The outcomes of gaining digital skills for young people's lives and wellbeing: a systematic evidence review. New Media Soc. https://doi.org/10.1177/14614448211043189
Marshall A (2021) Women's pathways to digital inclusion through digital labour in rural farming households. Aust Fem Stud 36(107):43–64
Marshall A, Dale A, Babacan H, Dezuanni M (2019) Connectivity and digital inclusion in far North Queensland's agricultural communities: policy-focused report. The Cairns Institute, James Cook University, Cairns. https://apo.org.au/node/253896
Marshall A, Dezuanni M, Burgess J, Thomas J, Wilson CK (2020) Australian farmers left behind in the digital economy: insights from the Australian Digital Inclusion Index. J Rural Stud 80:195–210
Marshall A, Wilson C-A, Dale A (2023) Telecommunications and natural disasters in rural Australia: The role of digital capability in building disaster resilience. Journal of Rural Studies, 100, Article number: 102996. https://doi.org/10.1016/j.jrurstud.2023.03.004

McKenzie B (2018) Australian Government response to the 2018 Regional Telecommunications Independent Committee report: 2018 regional telecommunications review: getting it right out there. Online: Australian Government. https://www.infrastructure.gov.au/media-centre/publications/australian-government-response-2018-regional-telecommunications-independent-committee-report-2018
Meat and Livestock Australia (MLA) Digital agriculture. https://www.mla.com.au/research-and-development/digital-agriculture/
Mills K (2010) A review of the "digital turn" in the new literacy studies. Rev Educ Res 80:246–271
NBN Co Ltd (2019) The technology that connects your premises. https://www1.nbnco.com.au/residential/learn/network-technology
Park S (2017) Digital inequalities in rural Australia: a double jeopardy of remoteness and social exclusion. J Rural Stud 54:399–407
Queensland Government (2018) Queensland agriculture snapshot 2018. Department of Agriculture and Fisheries, Brisbane. https://www.daf.qld.gov.au/__data/assets/pdf_file/0007/1383928/State-of-Agriculture-Report.pdf
Radovanović D, Holst C, Belur SB, Srivastava R, Houngbonon GV, Le Quentrec E et al (2020) Digital literacy key performance indicators for sustainable development. Soc Incl 8(2):151–167
Ragnedda M, Bruce M (2018) Introduction. In: Ragnedda M, Bruce M (eds) Digital inclusion: an international comparative analysis. Rowman & Littlefield, Lanham, pp vii–vxx
Rudd K, Conroy S, Tanner L (2007) New directions for communications: a broadband future for Australia—building a national broadband network. Australian Labor Party, Parliament of Australia, Canberra. Available at https://parlinfo.aph.gov.au/parlInfo/download/library/partypol/E2KM6/upload_binary/e2km65.pdf;fileType=application%2Fpdf#search=%22library/partypol/E2KM6%22
Salemink K, Strijker D, Bosworth G (2017) Rural development in the digital age: a systematic literature review on unequal ICT availability, adoption, and use in rural areas. J Rural Stud 54:360–371
Shiff D, McCluckey S, Somerset G, Eckemann R (2015) Regional telecommunications review 2015. Australian Government, Canberra. https://www.infrastructure.gov.au/sites/default/files/rtirc-independent-committee-review-2015-final.pdf
Simpson C (2017) Telstra 4G now covers 99 per cent of Australia's population. Gizomodo, August 1. https://www.gizmodo.com.au/2017/08/telstra-4g-now-covers-99-per-cent-of-australias-population/
Szoszkiewicz Ł (2018) Internet access as a new human right? State of the art on the threshold of 2020. Przegląd Prawniczy Uniwersytetu im Adama Mickiewicza 8(1):49–62
Thomas J, Barraket J, Wilson CK, Cook K, Louie YM, Holcombe-James I, Ewing S, MacDonald T (2018) Measuring Australia's digital divide: the Australian Digital Inclusion Index 2018. RMIT University, for Telstra, Melbourne
Thomas J, Barraket J, Wilson CK, Rennie E, Ewing S, MacDonald T (2019) Measuring Australia's digital divide: the Australian Digital Inclusion Index 2019. RMIT University and Swinburne University of Technology, for Telstra, Melbourne
Thomas J, Barraket J, Wilson CK, Holcombe-James I, Kennedy J, Rennie E, Ewing S, MacDonald T (2020) Measuring Australia's digital divide: the Australian Digital Inclusion Index 2020. RMIT and Swinburne University, Melbourne for Telstra
Thomas J, Barraket J, Parkinson S, Wilson CK, Holcombe-James I, Kennedy J, Mannell K, Brydon A (2021) Australian Digital Inclusion Index: 2021. RMIT, Swinburne University of Technology, and Telstra, Melbourne
Willis S, Tranter B (2006) Beyond the 'digital divide': internet diffusion and inequality in Australia. J Sociol 42(1):43–59

Chapter 11
Community Networks as Sustainable Infrastructure for Digital Skills

Raquel Rennó and Juliana Novaes

Introduction

Digital skills are part of the broader concept of digital literacy or digital *literacies,* as some authors prefer since it encompasses different approaches and solutions (Lankshear and Knobel 2008). Some authors define digital skills as capabilities one must master to obtain a particular result in the digital sphere, whereas digital literacy involves critical thinking skills when using the same tools (Bali 2016; Neumann 2017). It is possible to state that digital skills and digital literacy are connected and cannot be seen separately; digital skills compose some of the key measurable aspects of the broader concept of digital literacy. Both concepts are used in public policy planning for education and development on national and regional levels while also being studied by scholars in education, development, and technology areas; the variety of theoretical approaches and goals ended up providing a wide range of definitions of the terms showing some overlap but no consensus (Tinmaz et al. 2022).

A community network is a concept less present in academic research; it is a term that refers to different ways of developing and appropriating digital and communication technologies by local communities, from community radios to intranet networks, to villages connected to the Internet via optical fiber to mesh networks. Different technologies serve different needs, demands, and profiles; a

R. Rennó (✉)
PhD in Communications and Semiotics, Digital Programme Officer at Article19, member of International Center for Information Ethics (ICIE), Berlin, Germany
e-mail: raquelrenno@article19.org

J. Novaes
PhD Candidate in Transparency and Artificial Intelligence at the University of Leeds, Leeds, UK
e-mail: juliana.novaes.camargo@usp.br

D. Radovanović (ed.), *Digital Literacy and Inclusion*,
https://doi.org/10.1007/978-3-031-30808-6_11

community network in urban New York City differs significantly from one in a rural area in Brazil or an indigenous community in a remote area of Mexico—different human contexts also bring other funding models to these initiatives, generating a variety of results (Belli and Hadzic 2021).

Global partnerships between industry and international agencies target Latin America to increase digital literacy. Digital literacy remains a problem despite many of these initiatives mobilizing significant resources. The region presents a combination of low scores in international education rankings combined with a constant increase in the adoption of digital technologies for communication (International Telecommunication Union 2021; GSMA 2021). The International Telecommunication Union and the United Nations have tracked the development of connectivity worldwide and stated that low digital literacy is one of the main reasons for the current digital divide in developing countries (UNESCO 2018). According to the ITU's research (ITU 2020a), many people do not use the Internet or limit its use to chat apps because they do not know how to take advantage of what it can offer.

According to Radovanovic et al. (2020), there are three levels of the digital divide: the first level is the lack of Internet access or, as currently defined by the UN and the ITU, universal and meaningful connectivity (2022), the second level is the lack of digital literacies and competencies, and finally, the third level is the divide in life opportunities and benefits gained from the first two levels.

As a possible response to previously presented challenges, this chapter focused on *if* and *how* bottom-up solutions developed by local communities can offer a more suitable approach to overcome the digital divide in areas with limited to no Internet connectivity and low digital literacy and economic income. To achieve this goal, we carried out three case studies in different rural contexts in Brazil and Mexico, trying to find out lessons that could be learned from those initiatives.

To achieve the above-mentioned goals, we carried out three case studies in different rural contexts in Brazil and Mexico, trying to find out lessons that could be learned from those initiatives. These case studies were selected because they exemplify different scales and ways non-formal education projects connect with institutionalized sectors: *Art and Computation at Schools* have limited to no funding and rely on student scholarships when available. It is a case study about digital skills focused on basic programming and not Internet literacy skills (Timnaz et al. 2022); it was chosen because it represents a community-led alternative to leveraging people's appropriation of digital tools. *Community Provider Barra do* Açú is a community-led ISP aiming to become autonomous with a non-profit model based on Internet users' contributions. *Techio Communitario* is part of a network of programs and initiatives recently winning awards and grants and can keep its sustainability also by sharing its knowledge in courses for communities. In 2018, they were invited by the International Telecommunication Union (ITU), the multilateral agency responsible for managing and coordinating telecommunication on a global level, to offer specific courses on digital skills to Indigenous communities. All three case studies are located in rural or remote areas have a predominance of onsite training, and provide a bottom-up approach to digital skills learning in spaces where megaprojects from national and international commercial agreements have disrupted the environment and social tissue of local communities. Rather than rejecting digital technologies, these projects reappropriated them for the benefit of the community and proposed concrete alternatives for digital literacy training.

This chapter is organized as follows: section "Introduction" introduces discussed topics; section "Literature review" presents the literature review; section "Digital literacy in Latin America: where theory meets praxis" presents an overview of the leading digital literacy theories from Latin America; section "Methodology" introduces the methodology; section "Case studies" describes selected case studies in Brazil and Mexico, and section "Results" brings the paramount results; section "Discussion" presents the final discussion, and section "Limitations and future directions" provides the conclusions and some indications for future studies.

Literature Review

In order to map the research connecting community networks and digital literacy, a database survey was carried out on the most extensive academic research database in education, development, and technology. Articles, books, and chapters in English and Spanish on community networks and digital literacy were searched in the local and online University of Barcelona library repository; the Directory of Open Access Journals (DOAJ), JSTOR, SCOPUS, IEEE Xplore, Education Resources Information Center (ERIC), and Web of Science during the first half of August 2021.

The database search showed that digital literacy is briefly mentioned as one of the benefits of developing community networks in publications aimed at policymakers (APC and IDRC 2018). Still, practically no academic research focuses on the connection between these two concepts. From that point, it was decided the approach of the study presented in this chapter would be to analyze some of the digital literacy projects developed by the main actors in the field, public, and private sectors, and on the other hand, explore some cases studies of community networks with different constitutions and goals to see if and how digital literacy and skills are approached there.

Given the existing literature gaps, a literary review was conducted to understand digital literacy approaches by leading institutions that promote global digital literacy programs: international organizations, multilateral agencies, and the telecommunications industry.

International bodies and the industry often regard Information and Communication Technologies (ICTs) as an economic development tool. ICTs are a factor that would put countries ahead in the race for development. Reports published by the World Bank (2017), the Organization for Economic Co-operation and Development—OECD (2020), the United Nations (2018), and the Global System for Mobile Association—GSMA (2020) show the concept of ICTs and more specifically, digital technologies, associated with notions of competition between nations. This approach tends to reflect on ideas of digital literacy and skills as specific abilities required for the traditional labor market but often does not encompass broader skills, such as critical thinking and creativity. Moreover, this mainstream approach to digital literacy does not consider local communities' needs and knowledge systems.

The discourse of "being left behind" about digital technologies is frequent in Latin America and other Global South areas (Chan 2013). ICTs, particularly the Internet, are perceived as mere development tools and a straightforward way to shape the workforce to adapt to the market. The developmental approach to the Internet stems from its origins in telecommunication infrastructure. It stood out as a broader policy strategy in the early 2000s when the World Summit of Information Society (2003) coupled the concept of the *knowledge economy* with constant innovation through competitive skills. This concept was framed by the idea of Information and Communication Technologies for Development, or ICT4D (Heeks 2009). It was suggested that development via digital technologies could solve issues such as poverty, unemployment, and exclusion. It relates to the notion of *techno-solutionism*, where a few unique technological tools could be applied everywhere to solve different social and political problems (Chan 2013).

Educational projects initiated by the industry and international funding agencies, including the World Bank (2017), followed the same logic, framing digital literacy in a developmental narrative that includes it as part of a plan for countries to become economically competitive. UNESCO defined Digital literacy as "the ability to access, manage, understand, integrate, communicate, evaluate, and create information safely and appropriately through digital devices and networked technologies for participation in economic and social life (UNESCO 2018, p. 21). The private sector also measures digital literacy with specific digital skills, such as the GSMA's Digital Literacy Training Guide, which comprises a manual on using mobile technology resources for financial transactions (GSMA 2020).

Brown (2017) offers a detailed analysis of different digital skills proposed by international and multilateral agencies. The author concludes that, although it is essential to have clear indicators to measure progress in state actions and programs, too specific or narrow skills can become quickly outdated or irrelevant, challenging the achievement of meaningful and longstanding digital literacy results. It is in line with the OECD outcomes (2015) from interviews with professionals in innovative environments. The research aimed to understand which skills they think are the most useful, and the answers were: "coming up with new ideas and solutions," "a willingness to question ideas," and "the ability to present new ideas or products to an audience." Answers refer to critical and creative thinking rather than specific programs or software. Suppose digital skills indicators do not incorporate critical thinking, especially when measuring digital literacy in the Global South; in that case, there will always be a race toward the "digital future" that might seem near but impossible to achieve.

Moreover, it is not sustainable to train workers in the world's poorest regions to conform to the current system that has historically excluded and exploited them (Oxfam 2017). Some reports from multilateral agencies recognize the relevance of people's participation and the limits of one-size-fits-all solutions. The ITU (2020a, b) states in its report on sustainable connectivity (including digital literacy) that a participatory approach can be helpful in any situation. There are no fixed rules for determining each locality's actual or potential usage levels. Years before, the World Bank (2000) published a report stating that no development could be sustainable if the communities were not involved in the strategies. Radovanovic et al. (2020) highlight the discriminatory aspect in the framework used in high-income countries compared to what is used in middle- to low-income ones; while the first "incorporates multiliteracies/multimodal dimensions—which are the technical, cognitive, and social-emotional dimensions (...), most national frameworks in the low- and middle-income countries like India and Kenya conceptualize digital literacies only around the idea of competency and skill, which is too narrow a set of competencies(...)."

Digital Literacy in Latin America: Where Theory Meets Praxis

One of the issues with discriminatory standards for implementing digital literacy projects in the Global South is the misconception that there is no critical mass in these places to develop its own digital literacy process (Chan 2013). One of the most influential education theorists, Paulo Freire, and later, Mario Kaplún—to mention just a couple—focused on the use of media literacy and communication in the region in "a strict relationship between theory and practice, a strong critical and political commitment (...), and a profound rupture with the dominant positivism and functionalism in the emerging communication sciences." (Barranquero 2011).

Freire actively took part in the beginning of the leading political movements of the left in Brazil in the last years of military dictatorships, such as the Landless Movement (MST), the Worker's Party (PT), and the Central Workers Union Confederation (CUT). In most of these groups, Freire's popular education was implemented as a model for teaching and learning. Kaplún worked in the training of more than one hundred communicators for popular education, especially in rural areas, throughout 16 countries in Latin America (Kaplún 2002).

Freire proposed to teach using the context already known to the student rather than using outside information that can be too abstract (1968). This way of thinking critically about education and considering the immediate context when sharing knowledge has influenced a series of community-led initiatives. Most of the current community networks in the region inherited a particular approach that began with broadcasting media, particularly the community radios that have their roots in the 1940s. The 1960s and 1970s started the open spectrum activism in the region—the idea that the media should be available to all and not just a few actors authorized by the government (Freire and Guimarães 2012; Machado 1986), something many community networks in the regions still fight to achieve. The idea of community radios and community-led means of communication are directly aligned with this contextually focused approach and critical use of the media proposed by Freire (Freire and Guimarães 2012).

Methodology

The research on local communities' digital literacy processes started when one of the authors of this chapter worked together with computer scientists and digital media trainers to develop media and digital skills extension programs between 2013 and 2017. The targets were students from public schools, Landless Workers Movement members (MST), and Indigenous communities in rural areas of Brazil, aged between 13 and 18 years old. This research phase and an analysis of similar initiatives were published in Renno (2013; 2015; 2016). The participatory stage allowed to follow the development of the student's digital literacy, and there was the idea to study other cases to see how different contexts might generate similar positive results.

The three selected projects for the case studies were (i) a university extension program conducted in secondary schools in Northeast Brazil called *Art and Computation at Schools*; (ii) a community network in the state of Rio de Janeiro called *Community Provider Barra do Açú*, and (iii) a community radio and mobile communication initiative in Mexico called *Techio Comunitario*.

The case study of the extension program *Art and Computation at Schools* focused on the methods developed by project coordinator Jarbas Jácome, a computer scientist and lecturer from the *Universidade Federal do Recôncavo* da Bahia in the rural area of northeast Brazil. The case study was conducted via direct observation in the field between 2015 and 2017 while one of the authors was lecturing in the same area, following four in-training college students assisting Jácome at the school activities, and an in-depth video call interview with Jácome in February 2022.

The case study of the community provider Barra do *Açú* included an in-depth interview with Pedro Henrique Gomes Rodrigues, the project's Tech Coordinator, and Marcelo Saldanha from *IBEBrasil*. Both video call interviews were conducted in January 2022. For *Techio Comunitario* in Mexico, one of the project's coordinators, Carlos Baca-Feldman from Redes AC, was interviewed via video call in February 2022. The data was combined with desk research on the project's extensive documentation.

The 1.30 h in-depth interviews with coordinators of the three case studies aimed to understand how those projects started, which issues were addressed, and some of the outcomes achieved in digital skills. They were also asked about the interest, needs, and suitability of replicating the method in other contexts. The interviews were recorded via audio and transcribed; after the transcription, common elements from the plans and projects were highlighted so we could find the main concepts that would allow a comparison between the three cases (see Table 11.1) and list the replicability opportunities and challenges mentioned by the respondents. The method used to analyze the interviews was discourse analysis. This approach was

Table 11.1 Common attributes in the case studies

	Computer art at schools	Community provider Barra do Açú	Techio comunitario
Project started/ replicated by the community	Y: Trained students later become trainers	Y: Self-proposed project	Y: Self-proposed project
The community decides the project goal(s)	N: The project is proposed by a teacher to achieve a learning goal	Y: The project solves a community need	Y: The project solves a community need
Training process adaptable to context changes	yes	Yes	Yes
Projects focus on digital divide levels 1, 2, or 3	Levels 2 and 3	Levels 1, 2 and 3	Levels 1, 2 and 3
Tech learning by project	Y: Gamified projects for students	Y: Connect communities	Y: Connect communities
Tech skills learned return to the community	Y: Provide math and programming skills	Y: Self-sufficient community network	Y: Self-sufficient community radio/ network

chosen based on what is outlined by Tonkiss (2004), allowing the researcher to showcase the most relevant themes that appear in the sources. After the transcription, common elements from the plans and projects were highlighted so we could find the main concepts that would allow a comparison between the three cases (see Table 11.1). The process in this case involved organizing, pruning, and filtering information to identify possible categories for the data. The information was grouped into themes using the inductive method. The codes were grouped so we could find the main concepts that would allow a comparison between the three cases (see Table 11.1) and list the replicability, opportunities, and challenges mentioned by the respondents.

The case studies present evidence from small self-organized initiatives in Latin America that aim to overcome various levels of the digital divide using their own method and have no ties to international digital divide projects that have been implemented in different countries of the Global South for many decades. Although there is no official map of these initiatives due to reasons that go from their informal aspect to their choice to remain undetected in areas where there are direct conflicts with farmers and the local leadership, we know they exist all over the continent. Nevertheless, due to time limitations, we could not visit initiatives in other countries and interview the involved community. Our choice was to focus on small case studies that showed a variety of learning methods and development stages to see *if* and *how* they achieved their digital literacy goals.

Case Studies

Community Provider Barra do Açú, Brazil

Established in 2019 (with delays due to the COVID-19 pandemic in 2020), this community network provides Internet access via radio waves to the Barra do Açú district in Rio de Janeiro state. Connection via radio waves is a technological alternative in many rural areas where fixed Internet is transmitted via Wi-Fi in the absence of cable connections. The district is in the *Açú Superport* region of Rio de Janeiro state. Spanning 90 km^2, the port is one of the largest in the world, created by the Chinese government. The port facilities have led to internal displacements among the local population, severely polluted the air and soil, and reduced ecological and hydrological social-economic tissue, among other issues (Ejolt n.d.; Phillips (2010, September 15)). New traffic infrastructure that prioritizes the flow to and from the superport has disconnected the district of Barra do Açú from the city center. This disconnection caused by the port increased the prices of products and services, including the Internet. The port promised economic development and job opportunities, but the only jobs available to locals were temporary construction work that did not pay sufficiently.

A few individuals from a group that started a community Internet provider in a nearby neighborhood proposed to help create the provider in Barra do Açú. From the beginning, the idea was to provide capacity building through technical training to the local community to run the provider themselves. The Internet provider has a non-profit model, where a higher number of paying users would lower the cost for everyone. One of the first results of community organizing was that they started to feel more potent as a group and could negotiate with the public and private sectors on their terms. The community itself produces and manages the equipment and technology. The Internet provider organizers offer training on essential network management and software programming. The activity attracts people from the district as few vocational training opportunities are locally available.

The flexible operation of the group also allows members to change roles and try new tasks if desired, making the technical training more horizontal and diverse. The group recycles and uses DIY techniques, including building their transmission tower and recycling an old one with the help of the local blacksmith. Not only is the project replicable (it will be the second community provider in the area), but it is also part of a network of different community networks implemented in several small rural and remote communities across the country with the support of local NGOs. The *Barra do Açú Community Provider* is currently supported by NGO IBEBrasil, one of the most active in the field, and the local Federal Technical School with the idea of becoming financially autonomous soon with their non-profit model.

Techio Comunitario, Mexico

Redes AC is a collective that works in different areas: promoting small traditional producers, social tourism, urban agriculture, and participatory community development. Following Freire's ideas, it includes the right to media communication and autonomous technologies, so the collective offers training in telecommunications for Indigenous people based on their own choices and needs, including community radio, intranet, and community Internet service provider.

Baca-Feldman, interviewed for this research, is one of the coordinators of *Techio Comunitario,* a project focused on training Indigenous people to develop local digital media technologies. The word *tequio* is a Mexican term referring to the community effort to build something; the adaptation to *techio* connects with technology. The first phase of *Techio Comunitario's* work or "generation," as the organizers refer to it, occurred in 2006, and the second in 2012. It started with courses on the basics of digital technology. Later, the group created a specific course for community radio. The next step currently in progress is the development of community mobile communication providers. The project evolved into a network of local experts to share knowledge across different projects. In collaboration with many organizations and communities, they searched for ways to transform technologies and use them based on community principles and expertise to mitigate the risks they might entail. It is the way to weave a path towards technological autonomy collectively.

For the communities they worked with, connectivity is not an end in itself; "it is about implementing a communication project that helps to strengthen identity, autonomy, defense of territory and life." It starts by identifying a technological model contributing to achieving community goals and dreams.

In 2018, the ITU invited *Techio Comunitario* to develop an online on-site course on the ITU academy platform titled "Training technical promoters in Indigenous communities for the generation, development, and maintenance of communication and broadcasting network technologies." The course is currently supported by the ITU and the Fund for the Development of the Indigenous People of Latin America and the Caribbean (FILAC), ISOC, Frida, NIC Mexico, and *Rhizomatica.* It gathers people from 12 Latin American countries in the region each year. The course shows how the proposed method can be replicated and adapted to include participants from different countries in a hybrid format.

Techio Comunitario's successful work with the ITU proves that high-level agencies can develop partnerships with local initiatives and community-led efforts. Context-based education, respecting the communities' differences and needs, does not need to be at odds with the global step towards improving digital literacy.

Computer Art at Schools (Arte e computação nas escolas), Brazil

This digital literacy project was developed in a rural area of Bahia, where 96% of the population is of African descent, with around 15% of inhabitants in situations of extreme poverty and 80% at risk of poverty (MDA 2015). The oil industry has had an environmental and social impact on the greater area and negatively affected the local economy and society in the last decades, concentrating jobs in the oil processing area, bringing experts from outside the community while damaging the ecosystem where local communities used to live from (Fonseca 2007; Brito 2008). Lecturer and computer scientist Jarbas Jácome started the project Computer Art at Schools from 2011 to 2020 to teach programming logic to local schools. He noticed the basic math level among students was deficient and developed a method to make math concepts more applicable to students aged between 15 and 17 via the development of digital skills.

Jácome developed methods for teaching computational skills offline since there was no Internet access in the school (Jácome 2013). Together with his group of undergraduate students who assisted him in the project, he realized that the participants were more involved in activities that started with hardware. According to Jácome's experience, "if we start from basic electronics, then move to physical computing, and only then to actual software, we have a more engaging process, slowly raising the learning curve that is needed for harder programming." Students started by understanding what is a computer *bit* by touching an object that represents and acts like one crafted with their own hands, welding wires to produce sound, or watching the *bit* turn on a LED light.

A severe problem that affected the project was the cut of monthly support for students in 2020 and the pandemic. Individual financial support was the only possible way to participate in the project for many of them due to their low socioeconomic backgrounds, otherwise, they had to find unskilled jobs to support their families.

The lack of financial support is part of the country's general cut in education funding during the 2018–2022 federal policies. In terms of concrete results for the involved students, it was clear the project's outcome was very positive: students learned math by also understanding the basics of programming logic. The project provided digital literacy competencies and professional opportunities for the participants who managed to be financially supported throughout the learning process. Some students and local community members engaged in the project by becoming trainers. The trainers could replicate, adapt, and change the methodology in other places in the region after they finish school and college studies.

Results

The case study of *Community Provider Barra do Açú* addresses the first level of the digital divide, according to Radovanovic et al. (2020), aiming to bring connectivity to a community on its terms in a project managed by the same community. *Techio Communitario* addresses levels two and three, creating different communication channels and generating technological and media autonomy for the Indigenous communities. *Art and Computation at schools* also focused on levels two and three, teaching math and programming while allowing some students to become project replicators in other schools.

All three projects started from local ideas and methods directly influenced by popular education, especially Freire's critical pedagogy according to the interviews with the coordinators. The projects were developed to respond to the community's needs and demands, and the positive outcomes show how communities and perfectly capable of developing and replicating their own methods.

In all case studies, there are common elements (Table 11.1):

- The initiative starts with a small group meeting frequently onsite; everyone shares cultural codes and languages, making it easier to develop trust networks in the community of practice.
- There is no separation between skill learning and critical thinking. All three case studies start with a general proposition, but the end result comes from a joint decision of the group and reflects their needs or inputs; they know *why* they are doing the project, and the trainer guides them through the skills needed to achieve the goal;
- The tradition of popular education and media training are entangled;
- No method proposed in the three case studies is rigid: changes in the group and the context might require the training processes to be re-adapted.

There are some challenges to the projects, as well. At least in the first years of implementation, the financial support makes the projects possible. These communities in rural and remote areas do not have access to the economic system urban areas tend to offer; schools face inconsistent funding; youth migrate to bigger cities or work in jobs with low qualifications before finishing school. Scholarships and financial support can keep the project alive; in the case of community providers and community radios, it requires small financial support before it becomes sustainable and creates its funding mechanism.

Discussion

The case studies analysis in this chapter showed how community-led practices allow a broader, context-based, and adaptable understanding of digital skills and technology. The research showed solutions for different digital divide levels using a collaborative and critical approach (Bidwell 2021). It shows middle- and low-income countries can adopt indicators that prioritize critical and creative thinking and capacity to solve problems instead of limiting it to simply mastering the software or a tool in a specific digital environment (Stanley 2003). Consequently, educational goals can become more ambitious. Still, it will also allow partnerships and collaborations with other initiatives if indicators are kept flexible enough to respect their methodologies. The idea is not to ignore the future but allow local communities to create bridges between the past, the present, and the future, finding ways to connect to digital technologies that are meaningful to them; this can allow the learning process to keep going and avoid outdating knowledge that often happens when the learning goal is focused on just a technical task (Baca-Feldman and Hinojosa 2020).

The case studies also showed that, although economic development is part of the problem that has led to low digital literacy in the region, focusing only on the financial limitations does not help explain the whole situation nor come up with practical solutions (Hardy et al. 2018). If digital literacy strategies respect the people's context and include knowledge available among the community, they can become long-term solutions.

Educators and trainers in the three case studies understood the media's role in everyday lives and used it as part of strategies to learn and improve the quality of life in remote and rural areas. Popular educators in the 1960s–1980s saw a context with low general literacy and high media use; teaching using the media, which combined the oral and visual language, was a way to bring the written language closer to people (Mateus and Quiroz 2017).

In 2022, we are facing a similar situation: high mobile Internet use in the region—73% mobile Internet users in 2021 and growing to about 80% in 2025, with half of the growth coming from Brazil and Mexico. (GSMA 2021) combined with low digital skills—both Brazil and Mexico have around 20% of the population with just basic digital skills. (ITU, 2020a). By ignoring this recent history, digital literacy projects outside the area lose a great chance to build on previous knowledge that would allow projects to succeed and the communities to develop digital skills with a critical and sustainable foundation.

Global agencies and local initiatives collaborations may seem antithetical initially, but embracing differences as an opportunity instead of obstacles can be the needed solution. Multilateral agencies have already produced or financed enough research from experts from around the world, including Latin America (UNESCO 1998), showing the way to a digital literacy approach that embraces complexity and critical thinking. With hands-on training, such as the partnership between *Techio Comunitario* and the ITU, there can be a clear path toward improving society's digital and media literacy in rural and indigenous communities.

In the case of *Barra do Açú Community Provider*, the project is already the second community provider in the area; it is also part of a network of different community networks implemented in several small rural and remote communities across the country becoming financially autonomous with their non-profit model. Other providers from the network have different financing models depending on their location and community characteristics and the regulatory and financial support they manage to get from their local governments and civil and technical society.

On a more local level, even with the current financial constraints that put the *Art and Computation at Schools* on hold in the specific area where it was first implemented, the educators manage to share their ideas to be appropriated, adapted, and changed by other educators and digital artists working in an educational environment. The offline, DYI, and low-cost approach allowed it to be flexible enough to be widely shared among the wider academic community, including international ones, so the project continued in other schools.

Conclusion and Implications of the Study

Analyzing community networks under the digital literacy scope provides new light on the initiatives, limiting the risk of being restricted to its technical aspects and techno-cultural approaches, showing it can be a viable and effective bottom-up solution to overcome the lack of digital skills in areas with extreme social economic limitations. The multiple case studies analysis allowed the researchers to understand how diverse the learning methods and outcomes appear under the umbrella of digital literacy, confirming that the one-size-fits-all approach strategies and frameworks are not only insufficient but not suitable for understanding learning processes that are already taking place in different regions of the world.

In practical terms, this study implicates a broader understanding of how community practices could offer suitable answers to long-lasting digital literacy issues in rural areas with lower socioeconomic profiles beyond Internet access; it connects learning processes traditionally associated with female work with critical communication and digital technologies. When community networks are seen under the framework of digital literacy, they can become more than a low-cost alternative to Internet access and become a space for ongoing knowledge sharing in the digital field.

Limitations and Future Directions

The study was limited to three case studies and two countries in Latin America. Since there is a gap in academic studies evaluating the digital literacy potential of community networks, more studies should be carried out to map out more ways in which these initiatives develop their learning-by-doing methods, starting in Latin America, where the influence of critical pedagogy can be found and tracked in the communities, especially among community networks initiatives. It became clear that digital skills are developed at a broader level, not restricted to using a few tools or apps, allowing meaningful literacy to be achieved.

Future studies could cover other local initiatives, especially in different countries and regions. This can enrich the findings and add elements to sustainability solutions of community networks. Later, with a proper analysis corpus, mixed methods could be applied to find topics that would improve the assessment framework of digital literacy in the Global South, not limiting it to developmental or economic aspects but rather a multidisciplinary approach.

Acknowledgments The authors thank Don Le and Mehwish Ansari for their comments and insightful suggestions and the respondents for kindly sharing their experience.

References

Association for Progressive Communication and International Development Research Centre (2018) Global Information Society Watch – Community Networks. APC and IDRC. https://giswatch.org/sites/default/files/giswatch18_web_0.pdf
Baca-Feldman C, and Hinojosa D (2020) Propuestas metodologicas para disenar e implementar proyectos de comunicacion comunitaria. https://www.redesac.org.mx/_files/ugd/68af39_f3ccc6549625453ba8ef88a00b69f558.pdf
Bali M (2016) Knowing the difference between digital skills and digital literacies, and teaching both. https://www.literacyworldwide.org/blog/literacy-now/2016/02/03/knowing-the-difference-between-digital-skills-and-digital-literacies-and-teaching-both. Accessed 2 Sept 2022
Barranquero A (2011) Rediscovering the Latin American roots of participatory communication for social change, vol 8. Westminster Papers in Communication and Culture, p 154. https://doi.org/10.16997/wpcc.179
Belli L, Hadzic S (2021) Community networks: towards sustainable funding models. FGV Direito Rio, Rio de Janeiro. https://direitorio.fgv.br/conhecimento/community-networks-towards-sustainable-funding-models. Accessed 2 Sept 2022
Bidwell NJ (2021) Rural uncommoning: women, community networks and the enclosure of life. *ACM Trans Comput-Hum Interact 28*(3):19–50. https://doi.org/10.1145/3445793
Brito C (2008) A Petrobrás e a gestão do território no Recôncavo Baiano. EDUFBA. https://doi.org/10.7476/9788523209216
Brown M (2017) A critical review of frameworks for digital literacy: beyond the flashy, flimsy and faddish – part 13. *ASCILITE TELall Blog*. https://blog.ascilite.org/critical-review-of-frameworks-for-digital-literacy-beyond-the-flashy-flimsy-and-faddish-part-3/. Accessed 2 Sept 2022
Chan AS (2013) Networking Peripheries: Technological futures and the myth of digital universalism. MIT Press
EJOLT (n.d.) EJAtlas | mapping environmental justice. Environmental Justice Atlas. https://ejatlas.org/. Accessed 2 Sept 2022
Fonseca ÁCNdeO. (2007). Aspectos do desenvolvimiento regional no recôncavo sul baiano. https://www.semanticscholar.org/paper/Aspectos-do-Desenvolvimiento-Regional-no-Rec%C3%B4ncavo-Fonseca/a52df74db30a199646e27a922982b96147239780. Accessed 2 Sept 2022
Freire P (1968) Pedagogia do Oprimido. Record, Rio de Janeiro
Freire P, Guimarães S (2012) Educar Com a Mídia. Paz e Terra, São Paulo
Global System for Mobile Association (2020) Digital literacy training guide: a guide for mobile money agents and digital literacy change agents. https://www.gsma.com/mobilefordevelopment/wp-content/uploads/2020/11/Digital-Literacy-Training-Guide.pdf

Global System for Mobile Association (2021) The mobile economy: Latin America. Global System for Mobile Association. https://www.gsma.com/mobileeconomy/wp-content/uploads/2021/11/GSMA_ME_LATAM_2021.pdf

Hardy J, Dailey D, Wyche S, Makoto N (2018) Rural Computing: Beyond Access and Infrastructure. In: Companion of the 2018 ACM Conference on Computer Supported Cooperative Work and Social Computing (CSCW '18). Association for Computing Machinery, New York, pp 463–470. https://doi-org.sire.ub.edu/10.1145/3272973.3273008

Heeks R (2009, October 29) The ICT4D 2.0 manifesto: where next for ICTs and International Development? SSRN Scholarly Paper, Rochester, NY. https://doi.org/10.2139/ssrn.3477369

International Telecommunication Union (2020a) Digital skills insights 2020 | ITU academy. International Telecommunication Union. https://academy.itu.int/itu-d/projects-activities/research-publications/digital-skills-insights/digital-skills-insights-2020. Accessed 2 Sept 2022

International Telecommunication Union (2020b) ITU, World Telecommunication/ICT Indicators Databa. ITU. https://www.itu.int:443/en/ITU-D/Statistics/Dashboards/Pages/Digital-Development.aspx. Accessed 2 Sept 2022

International Telecommunication Union (2021) Facts and figures 2021. International Telecommunication Union. https://www.itu.int/itu-d/reports/statistics/facts-figures-2021/

Jacome J (2013) Arte e Computacao nas Escolas. https://jarbasjacome.files.wordpress.com/2013/12/metodologias_utilizadas_v2.pdf

Kaplún M (2002) Una pedagogía de la comunicación: (el comunicador popular) ([Ed. rev.].). La Habana: Editorial Caminos

Lankshear C, Knobel M (eds) (2008) Digital literacies: concepts, policies and practices. Peter Lang, New York

Machado A, Magri C, Masagão M, Guattari F (1986) Rádios livres: a reforma agrária no ar. Brasiliense, São Paulo

Mateus J-C, Quiroz Velasco MT (2017) Educommunication: a theoretical approach of studying media in school environments. Universidad de Lima. https://repositorio.ulima.edu.pe/handle/20.500.12724/4883. Accessed 2 Sept 2022

Ministério do Desenvolvimento Agrário (MDA) (2015) Perfil territorial Recôncavo Bahia, desenvolvimento territorial Sistema de Informações Territoriais. http://sit.mda.gov.br/download/caderno/caderno_territorial_187_Rec%C3%83%C2%B4ncavo%20-%20BA.pdf Accessed 4 Sept 2022

Neumann MM, Finger G, Neumann DL (2017) A conceptual framework for emergent digital literacy. Early Childhood Educ J 45(4):471–479. https://doi.org/10.1007/s10643-016-0792-z

OECD (2015) The innovation imperative: contributing to productivity, growth and well-being. OECD. https://doi.org/10.1787/9789264239814-en

OECD (2020) PISA 2018 results (volume VI): are students ready to thrive in an interconnected world? OECD. https://doi.org/10.1787/d5f68679-en

Oxfam International (2017, Jan 17) An economy for the 99%. Oxfam International. https://www.oxfam.org/en/research/economy-99. Accessed 2 Sept 2022

Phillips T (2010, Sept 15) Brazil's huge new port highlights China's drive into South America. The Guardian. https://www.theguardian.com/world/2010/sep/15/brazil-port-china-drive. Accessed 2 Sept 2022

Radovanović D, Holst C, Belur SB, Srivastava R, Houngbonon GV, Quentrec EL et al (2020) Digital literacy key performance indicators for sustainable development. Social Inclusion 8(2):151–167. https://doi.org/10.17645/si.v8i2.2587

Rennó R (2013) Aprender en las fronteras (o nadie educa a nadie): relaciones entre arte, ciencia y tecnología. Revista de Estudios de Juventud, 102. https://www.academia.edu/5708257/Aprender_en_las_fronteras_o_nadie_educa_a_nadie_relaciones_entre_arte_ciencia_y_tecnolog%C3%ADa. Accessed 2 Sept 2022

Rennó R (2015) Novos estudantes na velha sala de aula: o ensino da arte e tecnologia, entre institucionalização e mundos possíveis. Em Aberto 28(94). https://doi.org/10.24109/2176-6673.emaberto.28i94.1679

Rennó R (2016) Pinpinlab: Experimentações em arte, ciência e tecnologia DIY. LINKLIVRE. https://www.academia.edu/28835429/Pinpinlab_Experimenta%C3%A7%C3%B5es_em_arte_ci%C3%AAncia_e_tecnologia_DIY. Accessed 2 Sept 2022
Stanley LD (2003) Beyond access: psychosocial barriers to computer literacy special issue: ICTs and community networking. Inf Soc 19(5):407–416. https://doi.org/10.1080/715720560
Tinmaz H, Lee Y-T, Fanea-Ivanovici M, Baber H (2022) A systematic review on digital literacy. Smart Learn Environ 9(1):21. https://doi.org/10.1186/s40561-022-00204-y
Tonkiss F (2004) Analysing discourse. In: Seale C (ed) Researching society and culture. Sage, London, pp 245–260
United Nations Educational, Scientific and Cultural Organization (UNESCO) (1998) UNESCO and the development of education in Latin America and the Caribbean. In: Major Project of Education in Latin America and the Caribbean: bulletin, 45. OREALC, Santiago de Chile, pp 5–18. https://unesdoc.unesco.org/ark:/48223/pf0000262549. Accessed 4 Sept 2022
United Nations Educational, Scientific and Cultural Organization (UNESCO) (2018) Global framework of reference on digital literacy skills for indicator 4.4.2: Percentage of youth/adults who have achieved at least a minimum level of proficiency in digital literacy skill (Draft Report). Paris. http://uis.unesco.org/sites/default/files/documents/draft- report-global-framework-reference-digital-literacy- skills-indicator-4.4.2.pdf
World Bank (2000) Voices of the Poor. Can Anyone hear us? Narayan, D. World Bank. Oxford University Press. https://documents1.worldbank.org/curated/en/131441468779067441/pdf/multi0page.p
World Bank (2017) World Bank warns of "learning crisis" in global education. World Bank. https://www.worldbank.org/en/news/press-release/2017/09/26/world-bank-warns-of-learning-crisis-in-global-education. Accessed 2 Sept 2022

Chapter 12
Digital Inclusion Interventions for Digital Skills Education: Evaluating the Outcomes in Semi-Urban Communities in South Africa

Natasha Katunga, Carlynn Keating, Leona Craffert, and Leo Van Audenhove

Introduction

Grappling with the challenges inherent in being the most economically unequal country in the world, the South African government has underlined that 'all South Africans must benefit from the ability of the information and communications technology (ICT) sector to facilitate social development and improve the quality of life for individuals and communities' (Research ICT Africa 2020, p. 11). The magnitude of this vision is reflected in some of the socio-economic and technological realities of the country. South Africa is typically characterised as a middle-income country with a dual economy. Close to 57% of the 60.1 million population live below the country's poverty line, while the unemployment rate is at a staggering 34.4%—measured at the height of the global pandemic (World Bank 2021).

Despite the dominant view and aspiration to capitalise on the affordances of the transformative technologies of the Fourth Industrial Revolution (4IR) as building blocks of a progressive and prosperous society, unsettling inequalities prevail within the national ICT landscape. The most recent Network Readiness Index positions South Africa at number 70 of 130 participating countries (Dutta and Lanvin 2021). Highly advanced technology infrastructure is typically concentrated around the bigger cities and metropolitan areas, with limited access to it in rural and/or remote areas of the country (Statistics South Africa [StatsSA] 2022). The digital divide is a

N. Katunga (✉) · C. Keating · L. Craffert
CoLab for eInclusion and Social Innovation, University of the Western Cape, Bellville, South Africa
e-mail: nkatunga@uwc.ac.za

L. Van Audenhove
CoLab for eInclusion and Social Innovation, University of the Western Cape, Bellville, South Africa

iMEC-SMIT-Vrije Universiteit Brussel, Brussels, Belgium

D. Radovanović (ed.), *Digital Literacy and Inclusion*,
https://doi.org/10.1007/978-3-031-30808-6_12

stark reality in South Africa. Only 53% of the population had access to the Internet in 2017 (Gillwald et al. 2018),[1] with most gaining access through a mobile device (StatsSA 2022).

Various interventions have been implemented to address the country's persistent and widening digital divide. Of note are community organisations that have integrated digital inclusion offerings as part of their services. These organisations have been established through independent efforts, as well as configurations of collaboration between government, education, business and civil society and geared towards reaching the most vulnerable, under-resourced and digitally excluded communities.

Although not restricted to an economic focus, a primary objective of these organisations has been capacitating citizens to become more employable, entrepreneurial, expanding their skills and educational qualifications to improve their quality of life (Misuraca et al. 2014). These types of intermediaries[2] are identified in South Africa's national ICT policy and 4IR agenda as drivers of ICT awareness, access and digital skills development, particularly in under-resourced communities (URC) (Department of Communications and Digital Technologies [DCDT] 2020).

The skilling and training interventions of these organisations typically include digital skills training and alternative (formal and informal) learning options for people unable to afford traditional education institutions (Booi et al. 2019). They are often the sole gateway to these technologies and learning opportunities for many in these communities (Alao et al. 2017; Uys and Pather 2016). Focus is also on efforts to realise both feasible and effective approaches and methods to accelerate skills delivery for the South African context—specifically in reaching the most vulnerable and digitally excluded in society.

Despite the critical role of these organisations, there is limited evidence of the outcomes and influence of these (intermediaries') digital skills interventions in the lives of the intended beneficiaries (Avgerou 2010; Uys and Pather 2016, 2020). Traditionally, evaluation processes have been limited largely to outputs capturing the number of people who attended and/or completed training. While important, such output evidence is essentially an indication of 'volume rather than effectiveness' (Just Economics 2017). Surely this is insufficient to determine salient outcomes and effects over the longer term, as the foremost objective of digital skills interventions is the achievement of a meaningful influence on the lives of beneficiaries.

A necessary step is gaining evidence-based insight into the salient factors that contribute towards beneficial outcomes of digital skills interventions. In the absence of such informed understanding, we risk continuing to implement an unreflective or blanket approach to digital skills delivery, without sufficient contextualisation and consideration of the short-term and long-term benefits at the individual and community level.

[1] Limited information is available regarding Internet access at the individual level. National surveys typically focus on Internet access at the household level.

[2] In this chapter, the terms 'intermediary' and organisation are used interchangeably in reference to community organisations that are involved in digital inclusion interventions.

This chapter presents the findings of a quantitative survey study that sought to contribute to the practice of assessing the outcomes of digital skills training interventions. It provides insight into meaningful benefits derived from digital (mobile) literacy courses; salient factors contributing to such outcomes; and the application of methodological approaches and processes to evaluate the outcome of digital skills interventions in URC.

The remainder of this chapter: (a) provides a brief background and context to the subject of digital divides and evaluation of digital inclusion interventions; (b) discusses the building blocks of monitoring and evaluation (underpinning the research methodology); (c) explains the research methodology; (d) describes key research findings and (e) contextualises these findings in an integrated discussion and conclusion.

Literature Review

The Digital Divide: From Access to Outcomes

The digital divide, traditionally perceived as an issue of access, has been reframed to include focus on digital literacy (skills and competencies), usage and outcomes (Van Deursen and Van Dijk 2019; Helsper 2021). Three overarching digital divide levels are identified in the literature. The first-level divide is centred on the challenge of access. This goes beyond physical access to encompass quality, affordability, ubiquity and autonomy (freedom) aspects of ICT use (Van Deursen and Van Dijk 2019; Helsper 2021). The second-level divide relates to digital literacy and skills. This encompasses essential competencies that present-day citizens need to participate in a digital economy (Radovanović et al. 2020) and includes aspects of learning, problem-solving, critical thinking, creativity and self-regulation (Njenga 2018). In the current digital climate, digital literacy plays a crucial empowering and enabling role. On a practical level, what constitutes digital literacy is continuously evolving to adapt to changing requirements necessitated by rapid technological changes (Radovanović et al. 2020).

The third-level divide relates to improved livelihoods, benefits and outcomes, with current discourse asserting that inequalities (divides) in digital opportunities (access, skills and usage) contribute to inequalities in outcomes (Van Deursen and Van Dijk 2019; Helsper 2021). In essence, in a digital society, more advantaged citizens are systematically more likely to benefit, while those more disadvantaged are systematically less likely. Shifting focus from the first-level divide to include the second- and third-level divides enables us to understand the nuances of outcomes and how the affordances of ICT can become a reality.

It has become increasingly more important to determine and understand the outcome and impact of digital inclusion interventions (May and Barrantes 2015; Uys and Pather 2020). Common outcome themes include economic, social, cultural and personal well-being benefits (Helsper 2021). Several frameworks, models and theories exist that can generally be used to evaluate ICT for development interventions (Heeks and Molla 2009). For example, Sen's (1999) Capability Approach and Kleine's (2010) Choice Framework have dominated ICT for development literature. Focusing specifically on the evaluation of community digital skills training inventions, monitoring and evaluation (M&E) processes can also facilitate a holistic evaluation of the training intervention (International Telecommunication Union [ITU] et al. 2020)—which includes investigation of outcomes as well as the pathways through which beneficial outcomes are realised (Just Economics 2017).

Prominent examples of M&E frameworks used to evaluate outcomes and/or impact of digital inclusion interventions include the 'MIREIA e-Inclusion Intermediaries Impact Assessment Framework (MIREIA eI2-IAF)', designed with specific regard to interventions focusing on the use of ICT to enhance the employability of groups at risk of exclusion (Misuraca et al. 2014); and the United Kingdom's 'Digital Inclusion Evaluation Toolkit' (Just Economics 2017), designed to understand and share the results of the effectiveness and ability of digital inclusion interventions to meet local needs relating to significant economic, social and health benefits.

Monitoring and Evaluation: Theoretical Basis

Monitoring and evaluation assist in (i) defining and understanding intervention objectives; (ii) conceptualising the relationships between objectives; (iii) defining the underpinning activities required to achieve the stated objectives and (iv) describing the anticipated outcomes (World Health Organisation [WHO] 2016). The process is supported by an underlying framework which commonly includes the Theory of Change (ToC)—a comprehensive description of how and why the intervention is expected to achieve the intended objectives (Department of Planning Monitoring Evaluation [DPME] 2021)—and results chain and logical models which can be used to visually illustrate causal links of the ToC.

Key conceptual building blocks of the framework include *inputs*—context and resources required to undertake the intervention; *activities*—actions taken to deliver the intervention; *outputs*—direct results of activities; *outcomes*—expected (and unexpected) changes that are anticipated to occur because of the activities of the intervention; and *impact*—related to the long-term broad effects of the intervention for the target participants, the economy and society. Figure 12.1 presents these building blocks as reflected in a results chain.

Measurement indicators need to be adjusted to and aligned with the nature and scope of the digital skills intervention. They guide the identification of relevant contextual attributes (inputs and activities) and measure output, outcomes and impact (Just Economics 2017).

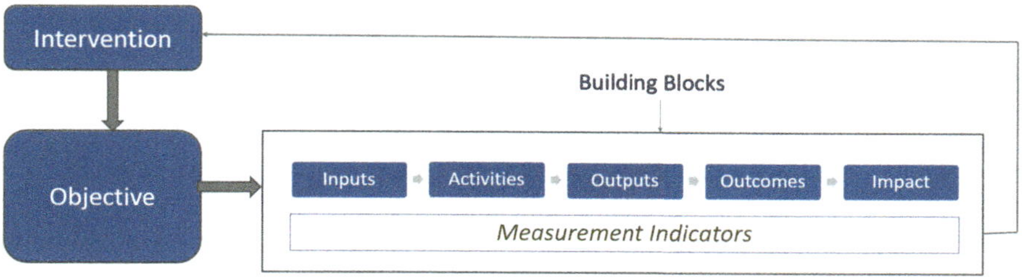

Fig. 12.1 Results chain model

Methodology of the Empirical Study

The following section presents the nature of the digital training intervention and the research approach.

Describing the Digital Training Intervention

Given the significant percentage of the population who rely predominantly on their mobile devices to harness the affordances of ICT for life and work, a Mobile Literacy course was developed to provide support to these citizens.[3] The course was structured to focus on digital (mobile) literacy competencies as defined in the Digital Skills Framework One (DSFOne), a digital skills competency framework tailored to the South African context (Claassen 2021).[4] The Mobile Literacy programme was developed to enable participants to master the digital literacy competencies, namely handling of information, communication and collaboration, safety and security, problem-solving and transacting. The 6–8 h course is designed for face-to-face or blended learning approaches and consists of a student guide, video clips to demonstrate learning activities (in three local languages) and presenter notes. After course refinement through pilot implementations and the upskilling of trainers, the course was implemented in 2020 through four community-based organisations in different peri-urban and rural environments in the Western Cape province of South Africa.

The clearly defined starting timeline, complexity of the training context (due to the Coronavirus disease [COVID-19]) and, therefore, the heightened urgency to ensure a beneficial outcome positioned the mobile skills intervention as an ideal scenario for assessing the outcome of skills interventions in URC.

[3]The course was developed by the CoLab for e-Inclusion and Social Innovation based at the University of the Western Cape and funded by the Department of Communications and Digital Technologies, through the National Electronic Media Institute of South Africa.

[4]See https://www.wcapecolab.org/dsf1

Developing a Results Chain

A results chain consisting of *input*, *activities*, *output*, *outcome* and *impact* building blocks and indicators specifically aligned to the objectives of the Mobile Literacy training intervention was developed to assess the outcomes of the intervention. Core principles of ToC and results chains aimed at evaluating digital inclusion interventions were used as a guide in the design of the Mobile Literacy training intervention results chain, which is outlined in Fig. 12.2.

For this study, *input* refers to the context and resources of the intermediary, the identification and recruitment of appropriate participants, the course content, and the skills levels of facilitators. Given the findings of a previous study in terms of the role of intermediaries in facilitating digital inclusion (Katunga 2019), it was deemed necessary to expand the input dimension to also include contextual information on intermediaries and their environments. *Activities* focus on the training delivery and support activities, while the number of attendants, successful completion and/or performance level typically relate to the *output* dimension. The *outcome* dimension refers to the short-term and mid- to longer-term outcomes or benefits of the programme as experienced by beneficiaries. Although the *impact* dimension forms part of the results chain, this study did not include impact as part of the Mobile Literacy evaluation process. The focus was on outcome benefits.

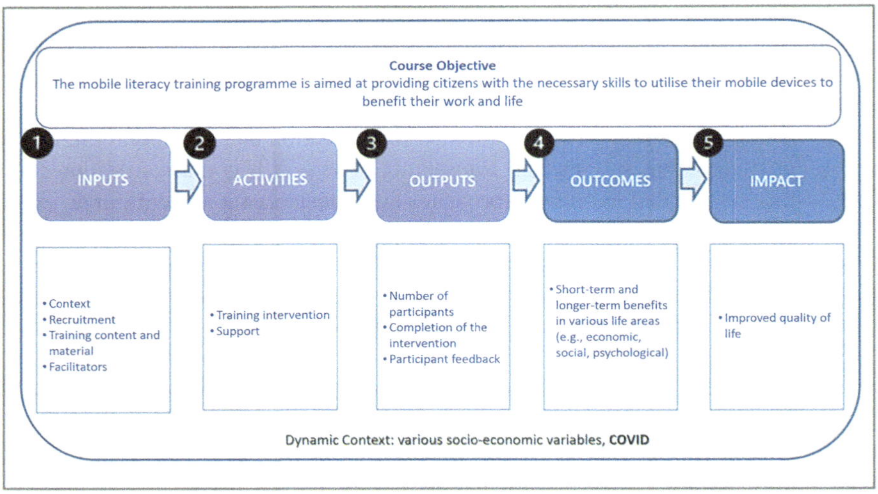

- Building blocks 1, 2 and 3: data collected from intermediaries - **Survey 1**
- Building block 4: data collected from beneficiaries - **Survey 2**

Fig. 12.2 Mobile Literacy training intervention results chain

Survey Instrument Development

To obtain the necessary information outlined in the results chain, a quantitative research approach was followed, applying survey methodology (Creswell and Creswell 2018). As the Mobile Literacy results chain (Fig. 12.2) includes the perspectives on two different units of analysis (training provider and beneficiaries), two separate surveys were constructed and implemented between November 2021 and February 2022 to engage with the respective groups.

Survey 1: Focus on intermediaries

The quick-scan methodology (Van Audenhove et al. 2016) was applied to obtain contextual information about the four training providers, focusing on the first three blocks of the results chain (input, activities, outputs). This methodology typically utilises a collection of structured questions (open and closed) to obtain an impression of areas and/or organisations where variance is suspected. The survey focused on gaining information on the available infrastructure where training interventions were conducted, the nature and scope of services, target audiences, training pedagogy and throughput.

Survey 2: Focus on beneficiaries

A survey consisting of open-ended and closed-ended questions was constructed to capture beneficiaries' experiences of the Mobile Literacy course and perceived benefits or outcomes of the course (building block 4 of the results chain). As such, the survey was not designed to test abilities gained or measure competence against a competency profile (or framework), but rather individual perceptions of meaningful benefits derived as a result of the training intervention.

Data Gathering Process

The quick-scan questionnaire (survey 1) was sent to intermediaries via e-mail for completion at their own convenience.

Using the information provided by intermediaries as their achieved output (number of beneficiaries trained) a database of 4040 beneficiaries was compiled. Some participants had to be omitted due to either being under the age of 18 (i.e., minors) or incomplete contact information, resulting in a total number of 3650 participants.

Given the reality of COVID-19, an online approach had to be adopted for the distribution of the beneficiaries' survey (survey 2). The link along with the necessary information and research consent form was sent to the e-mail addresses of the 3650 Mobile Literacy training beneficiaries and 3548 were delivered successfully.

A very low response rate was achieved, and the data-gathering process had to be adjusted to increase responses (Nulty 2008). Reminders, inclusive of the survey link were subsequently sent to the mobile devices of beneficiaries via short message service (SMS) text messages. In addition, incentives of 1 GB data were offered to the first 100 participants to submit their completed surveys.

As the data collection process via e-mail and SMS did not yield the desired results, the traditional fieldwork approach was adopted as a last resort. Representatives of the intermediaries were recruited as fieldworkers and trained to assist in the administration of the survey. They contributed towards the data collection process by sharing the survey information (original communication and survey link) through their normal communication and marketing channels, motivating beneficiaries to participate in the study. They supported beneficiaries with the completion of questionnaires by inviting them to their premises, providing them access to the Internet and devices and, in some cases, by printing hard copies for completion. Hard copies were scanned and the data captured.

Response Rate and Demographic Profile of Participants

Of the 3650 listed beneficiaries, 557 responded to the questionnaire. This number reduced to 510 after data cleaning. With a population size of 3650, a confidence level of 95% and a margin of error of 5%, the required sample size would be 348. In this case, the sample size was 510, which means the confidence level increased to about 98.5%. If the confidence level is kept at 95%, then the margin of error reduces to about 4%. Thus, with a sample size of 510 for the population of 3650, one does not always know that the correct answer has been found. However, we do know that there is a 98.5% chance that responses are within a 5% margin of error of the correct answer.

Of the total number of respondents who completed the evaluation ($N = 510$[5]), the majority (88%) were between the ages of 18 and 35, with 12% being 36 years and older. Sixty-eight percent (68%) of the respondents were female and 32% were male. Regarding race, of the 505 respondents to the survey, 28% were Black African, while 72% were coloured (of mixed race).

In terms of education, (of 509) 24% had a post-school qualification, 56% had completed high school, while the rest had either primary school education or no formal education. Focusing on employment, 498 responses were received, 11% of which were employed full time, 40% part time, 4% were self-employed and 31% were unemployed. The remaining respondents were either students, retired or did unpaid housework. Thus, about 55% of respondents reported some form of employment.

[5] It must be noted that the results are presented as a percentage of values received per variable (i.e. per question). Missing values were omitted, hence the inconsistency in the sample size per question.

Reporting of the Findings

The following section presents the findings of: (i) the quick-scan study focused on intermediaries (summarised in Table 12.1); and (ii) the survey focused on beneficiaries of the Mobile Literacy training intervention.

Table 12.1 Intermediary profiles as obtained from the quick-scan study

	Intermediary 1	Intermediary 2	Intermediary 3	Intermediary 4
Described as	Training institution Innovation hub	Training institution Multi-purpose community centre	Training institution	Training institution Multi-purpose community centre Innovation hub
Location type	Peri-urban	Urban	Peri-urban	Peri-urban
Footprint – where services are provided	Central premise Off-site venues	Central premise	Central premise Off-site venues	Central premise Off-site venues
Number of Mobile Literacy trainees	1247 [34% of training beneficiaries]	430 [12% of training beneficiaries]	1509 [41% of training beneficiaries]	464 [13% of training beneficiaries]
Venue ownership	Hires venues Access to free venues	Owns venues Access to free venues	Hires venues	Owns venues Hires venues Access to free venues
Available ICT infrastructure (Internet access, computers, mobile devices)	Internet access Own and hire computers	Internet access Own computers and mobile devices	Internet access Own and make use of other freely available computers and mobile devices	Uses Internet access of hired venues Makes use of other freely available computers and mobile devices
Services provided	Public access to computers and the Internet Training services (digital and other) Job seeking and CV writing support Small business support Participate in community development initiatives	Public access to computers and the Internet Training services (digital and other) Job seeking and CV writing support Small business support Participate in community development initiatives	Public access to computers and the Internet Training services (digital and other) Job seeking and CV writing support Small business support Participate in community development initiatives Facilitate community engagement	Training services (digital) Small business support Participate in community development initiatives Facilitate community engagement

(continued)

Table 12.1 (continued)

	Intermediary 1	Intermediary 2	Intermediary 3	Intermediary 4
Staff complement	Six to 10 people	Six to 10 people	More than 20 people	More than 20 people
Target groups	Low-skilled Low-income Unemployed youth Women Small businesses Students	Low-skilled Low-income Unemployed youth The elderly People with disabilities Women Small businesses	Unemployed youth People with disabilities Women Small businesses Students	Low-skilled Low-income Unemployed youth The elderly People with disabilities Women Small businesses Students
Cost of training	Most of the training courses require a fee; a few courses are free	Most of the training courses are free; a few courses require a fee	All training courses are free	All training courses are free
Training approach	Face-to-face Online Blended (face-to-face and online)	Face-to-face Online Blended (face-to-face and online)	Blended (face-to-face and online)	Face-to-face Blended (face-to-face and online)

Findings Related to Intermediaries

It is evident from the information obtained from the quick-scan study that the four intermediaries had several characteristics in common: they all operated from a fixed physical location, had access to the Internet, and either owned, hired or had access to computers and mobile devices. At a basic level, all the intermediaries regarded themselves as training institutions (although not necessarily confined to training), with citizens from vulnerable groupings or under-resourced contexts as the dominant target group. They offered a range of services, with training interventions, support for small businesses and community development initiatives as shared interests.

When the information on the intermediary profile is related to the input dimension of the evaluation framework, it is clear that the intermediaries had all the necessary physical and ICT-related infrastructure (venues, Internet access, devices) and resources (trainers, course content) at their disposal to deliver the Mobile Literacy skills intervention. Intermediaries had access to the course content (offered in three languages), were trained in course delivery and had several years of experience with digital skills intervention. Furthermore, as indicated in the profile above (Table 12.1), the Mobile Literacy skills development interventions were targeted predominantly at citizens who found themselves in precarious conditions (under-resourced environments), students, and the youth. Consequently, it seems fair to deduce that the participating intermediaries had the necessary input indicators at their disposal to perform the required activities that resulted in an output of 4040 citizens who successfully completed the Mobile Literacy course. This number refers to the typical volume count.

Findings Related to Beneficiaries of the Mobile Literacy Course

The following discussion reports on survey findings related to: (i) the digital inclusion profile of beneficiaries in terms of access to ICT; and (ii) the outcomes of the course as perceived by beneficiaries.

Digital Inclusion Profile of Beneficiaries

Of the 426 responses on smartphone ownership, 86% owned a smartphone, while 11% had access to one through either a friend or family member. Three per cent stated that they did not own or have access to a smartphone at home, school or their place of work. Regarding laptop computers, 407 responses were received, 31% of whom owned a laptop, 35% did not own a laptop but had access to one through either a friend or family member, and 34% did not own or have access to a laptop device at home, school or place of work. A total of 449 responses were received regarding Internet access, of which 92% had access through a mobile device. Of this group, 40% did so by buying mobile data, 26% had access to Wi-Fi at home, and the rest either made use of free Wi-Fi hotspots in public buildings like libraries and churches, or they used the Wi-Fi provided at work. Interestingly, 11% also made use of free Internet websites and applications like Facebook Lite.

Perceived Training Outcomes

It is clear that the majority of the research participants (87% of 445) were of the opinion that they benefited from attending the course. The value gained from the course is evident, given that 71% of (406) respondents had already recommended the course to someone else at the time of the survey, while 28% had not recommended the course but stated that they would. Supported by the open-ended responses regarding how and/or what respondents gained from the training, the benefits were divided into three overarching themes: (i) psychological, (ii) economic and (iii) social benefits.

Psychological Benefits

In the context of this study, psychological benefits encompass changes in behaviour regarding the use of mobile devices, gaining feelings of self-awareness, empowerment, motivation and confidence and changes in mindset and attitude regarding the value of technology. Of the 409 people who responded to this question, 84% either agreed or strongly agreed that they were using a mobile device for more work and personal purposes because of the training (8% disagreed and the rest were not sure). Furthermore, from 410 responses, 94% either agreed or strongly agreed that the training had made them more interested in exploring the Internet and other digital devices.

An objective of the course is to provide respondents with information about key concepts of mobile digital literacy and to open their minds to the 'bigger picture'. Answers to the open-ended questions show that respondents gained an understanding of technology in general, i.e. the 4IR and its influence on how people communicate, learn and work, and consequently its influence on the changing world of work. *'The experiences I gained from the course is* [sic] *good, because I have more knowledge about technology in and around the world. I have implemented those skills in my studies because I do study computer literacy'* (respondent 312).

Gaining knowledge and the ability to use a mobile device for more purposes gave some respondents a boost of confidence, even to apply for jobs. Eighty-eight per cent (88%) of 409 respondents either agreed or strongly agreed that the training helped them to become more confident in using mobile devices. Only 6% either disagreed or strongly disagreed, with the rest not being sure. Furthermore, knowing about the security risks associated with activities such as online banking, responding to unknown e-mails and sharing personal information and images influenced some respondents to change the way they use mobile devices to protect themselves.

Economic Benefits

Economic benefits in this regard relate to respondents gaining knowledge and the ability to use mobile devices to access employment opportunities, manage resumés, save money, facilitate business transactions and conduct financial transactions. Out of 400 responses, 67% stated that because of the training, they were using a mobile device to search and apply for job vacancies. The training also played a role in 62% of (398) respondents stating that they were using a mobile device for financial activities, for instance by using mobile banking applications.

Of 408 respondents, 89% stated that they had become more productive because of integrating mobile applications into activities they would have had to do manually. For example, they used their device to scan documents and e-mail them to people instead of going to a phone shop or Internet café and paying to have that done: *'Now I use scan on my device, no payment money to scan my documents and it works perfectly'* (respondent 314). In addition to being more productive, some respondents saved time and money: *'I benefited bcoz i don't have to travel that much if I want to get some forms like Z83 from the police station and any other documents, I can easily download from my smartphone'* (respondent 48).

Social Benefits

Social benefits entailed participants gaining feelings of inclusion, social capital, a desire to explore and learn more about technology and how it can be useful. From 395 responses, it was found that 62% had started using a mobile device for entertainment purposes (including playing games and watching videos) and that the training had played a role in this. Furthermore, 86% of (406) respondents stated that they had started using a mobile device to help carry out even mundane daily tasks and activities, which became easier or were completed much faster. For instance, they used Google Maps for directions: '*Being a full-time student and not always having the time to walk around with a laptop, doing my assignments on my mobile device makes my life so much easier after the training that was provided*' (respondent 7).

From 410 responses, 94% either agreed or strongly agreed that the training had made them more interested in exploring the Internet and other digital devices. This included new ways of interacting and communicating with people. For instance, 64% of 401 respondents had started using applications like social media, instant messaging (WhatsApp) and e-mail to communicate because of the training. Out of 398 responses, 62% had started joining different social media groups to interact with new people outside of their family and friends. A total of 62% of 398 respondents had also joined community WhatsApp groups to participate in discussions about community issues and events.

Respondents expressed increased feelings of inclusion, as they could participate in discussions about technology: '*I can also participate in discussions about the 4th industrial revolution ... it's very interesting for me*' (respondent 49). Other respondents felt more included in their children's school life because they were able to help with research, i.e. searching the Internet for information for their children's homework.

Discussion and Conclusion

COVID-19 necessitated the acceleration of digital skills development to facilitate citizens access to critical services and information for the purposes of societal inclusion. Given shrinking training budgets and restricted face-to-face interactions (due to the reallocation of funding to health-related projects, the national lockdown and social distancing measures), the Mobile Literacy course offered a viable option to facilitate the acquisition of basic but essential digital skills by citizens.

This study sought to gain insight into the outcomes of such digital skills delivery, in terms of meaningful benefits derived as perceived and reported by beneficiaries. Despite initial criticism, as there is a general assumption that such skills can be self-taught, it is encouraging that, despite the short duration of the Mobile Literacy course (6–8 h), the post-training evaluation points to clear perceived benefits related to psychological, economic and social dimensions. Some of the key outcomes are highlighted in Table 12.2.

The pathway to these benefits was evident with a clear sequential relationship between intervention *inputs*, *activities* and achieved *outcomes*. The fact that intermediaries (i) were well established in the communities and trusted, (ii) had access to up-to-date content, venues and ICT infrastructure and (iii) were skilled and capacitated to provide the training and to support participants were essential in attaining positive outcomes. These interrelated factors align with core measurement indicators of existing digital inclusion intervention evaluation frameworks (e.g. MIREIA e-Inclusion Intermediaries Impact Assessment Framework).

Table 12.2 Overview of outcomes for beneficiaries

Psychological	Economic	Social
Increased confidence to use mobile technology, Motivation to gain more advanced digital skills, Improved cybersecurity and safety awareness	Improved access to employment opportunities, Financial savings, Increased use of mobile technologies in professional activities, More efficient use of mobile technologies in financial activities	Extended use of mobile technologies in communication and social life, Use of mobile technologies for practical activities (e.g. assisting children with homework)

The prevailing assumption suggests that equalities in digital opportunities lead to equality in realised benefits. Although intermediaries provided access to technologies (level 1 of the digital divide) and facilitated skills development (level 2), it was evident that the degree and scope of benefits (level 3) were not equal among the beneficiaries—as was the case in digital literacy interventions in other low-income areas (Radovanović et al. 2020). This is in line with Helsper's (2021) model of socio-digital inequalities, which illustrates that addressing access, skills and use is not enough. We need to consider nuances at the individual level in terms of inequalities in social, economic, cultural and personal well-being, which consequently influence the equality of benefits. While this is undoubtedly complex, it is necessary to bear in mind if we are to determine why there is a spectrum of benefits.

Baseline assessments (focusing on the nature of first and second-level divides) should thus form part of the digital skills intervention assessment processes, included in the *inputs* dimension, that are done before the intervention activities. Only the individual can shed light on their gradations of exclusion, along with their associated challenges and opportunities and this must be considered in the activities of the intervention towards achieving the outcomes. Helsper (2021) emphasises that outcomes are not homogeneous, but subtle and inherent in a beneficiary's environment and context.

In terms of the administration of the Mobile Literacy post-training evaluation, key learning and observations were made. Obtaining the participation of training beneficiaries in view of ascertaining the perceived outcomes or benefits of the intervention posed significant challenges. Although necessitated by COVID-19, relying on online assessments was clearly not successful. Expanding the online survey (via e-mail) to sending SMS text messaging to beneficiaries' mobile numbers, yielded only a slight increase in survey responses. It transpired that research participation was hampered by the beneficiaries' lack of Internet access and/or high data costs, which made participation an expensive and even unaffordable exercise.[6] Accessing

[6] WhatsApp messaging was also explored for survey purposes. However, it requires the registration of a WhatsApp business service, which at that point in time was not within the policy framework of the university due to recent changes in legislation and regulation.

training beneficiaries via the communication network of intermediaries proved to be a more successful approach. Intermediaries supported research participants by providing access to the Internet and devices and, in some instances, acted as survey administrators, assisting beneficiaries in completing the questionnaire.

Following from the notion that inequalities in digital opportunities (first and second-level digital divide) lead to inequalities in terms of outcomes (or benefits) (Radovanović et al. 2020; Helsper 2021), it may be argued that this is equally true for the ability of training beneficiaries to participate in post-training evaluation. The profiles of beneficiaries in terms of access to ICT reveal that many rely on the infrastructure and support of intermediaries. The administration of an assessment process (data-gathering approach) in environments where the first- and second-level digital divide is still a reality (Scheerder et al. 2017) needs to take cognisance of the unequal digital circumstances of the beneficiaries. This should at a minimum include support in terms of providing Internet and device access at accessible venues, contributing to either data costs or travel expenses to venues and even the zero-rating of surveys. It is worth noting the critical supportive role of intermediaries which emerged from these findings, evident in both (i) the social support—the instrumental, informational and emotional aid received from support networks, assisting an individual's use of digital technologies (Asmar et al. 2020)—they provide in effectively executing digital inclusion interventions; and (ii) the support they provide in enabling evaluations of such interventions.

Findings related to the perceived outcomes of the Mobile Literacy training intervention are encouraging and point to tangible and intangible outcomes. Although a representative sample was achieved, an inherent bias in the sample should be noted. There is a likelihood that beneficiaries who had a positive experience of the training were more inclined to respond to the assessment than those who had a negative experience. Following the argument of Nulty (2008), this inherent bias can typically be addressed by applying multiple methods to assess perceived outcomes or benefits. Evaluation assessments should therefore ideally develop and apply multiple methods (measurement instruments and modes) to explore the different perspectives for a more informed understanding of the nuances in perceived outcomes.

Finally, the data collection process should ideally be constructed to consist of multiple approaches (i.e. online assessment, SMS, face-to-face) to ensure a higher response rate while careful consideration should be given to the sampling method.

Evaluation assessments are costly and time-consuming exercises. The proposed adjustments for the application of multiple assessment instruments and data-gathering approaches may have cost and time implications. However, to help facilitate and encourage the application of digital skills intervention assessments on a more regular basis and for a better understanding of the nuances related to the persistent digital divide, innovative approaches applicable specifically in the context of URC need to be developed and tested.

Acknowledgement This research project was financially supported by the National Electronic Media Institute of South Africa (NEMISA), a portfolio organisation of the South African National Department of Communications and Digital Technologies (DCDT) and the University of the Western Cape (UWC).[7]

References

Alao A, Lwaga TE, Chigona W (2017) Telecentres use in rural communities and women empowerment: case of Western Cape. IFIP Adv Inf Commun Technol 504:119–134

Asmar A, Van Audenhove L, Mariën I (2020) Social support for digital inclusion: towards a typology of social support patterns. Soc Incl 8(2):138. https://doi.org/10.17645/si.v8i2.2627

Avgerou C (2010) Discourses on ICT and development. Inf Technol Int Dev 6(3):1–18

Booi LS, Chigona W, Maliwichi P, Kunene K (2019) The influence of telecentres on the economic empowerment of the youth in disadvantaged communities of South Africa. In: Nielsen P, Kimaro HC (eds) Information and Communication Technologies for Development: Strengthening Southern-Driven Cooperation as a catalyst for ICT4D. Springer, Cham, pp 152–167. https://doi.org/10.1007/978-3-030-18400-1

Claassen W (2021) Digital Skills Framework One ("DSFOne"). Available at: https://www.wcape-colab.org/search/tags/dsfone

Creswell JW, Creswell JD (2018) Research design: qualitative, quantitative, and mixed methods approaches. SAGE, Thousand Oaks

Department of Communications and Digital Technologies [DCDT] (2020) National digital and future skills strategy. Government Gazette Issue 43730. https://www.gov.za/sites/default/files/gcis_document/202009/43730gen513.pdf

Department of Planning Monitoring Evaluation [DPME] (2021) Evidence management for an effective and efficient Socio-economic Impact Assessment System (SEIAS): an organizational guide on using evidence when implementing SEIAS. Pretoria. Available at: https://www.dpme.gov.za/publications/PolicyFramework/SEIAS Evidence Guide.pdf

Dutta S, Lanvin B (2021) The network readiness index 2021. Portulans Institute, Washington, DC. https://networkreadinessindex.org/wp-content/uploads/reports/nri_2021.pdf

Gillwald A, Mothobi O, Rademan B (2018) Policy paper no.5, series 5: after access. The state of ICT in South Africa. Research ICT Africa, Cape Town

Heeks R, Molla A (2009) Development informatics compendium of approaches, impact assessment of ICT-for-development projects: a compendium of approaches. 36. https://doi.org/10.2139/ssrn.3477380

Helsper EJ (2021) The digital disconnect: the social causes and consequences of digital inequalities. SAGE, London

International Telecommunication Union [ITU], United Nations Educational, Scientific and Cultural Organization [UNESCO] and United Nations International Children's Emergency Fund [UNICEF] (2020) The digital transformation of education: connecting school, empowering learners. Available at: https://www.broadbandcommission.org/wp-content/uploads/2021/02/WGSchoolConnectivity_report2020.pdf

Just Economics (2017) The digital inclusion evaluation toolkit: an overview. Government Digital Service, Cabinet Office, London

Katunga N (2019) Communicating for development using social media: a case study of e-inclusion intermediaries in under-resourced communities. Doctoral dissertation, University of the Western Cape and Vrije Universiteit Brussel. Available at: http://hdl.handle.net/11394/6999

[7] Our gratitude goes to the community-based organisations that assisted us in the completion of the beneficiary surveys.

Kleine D (2010) ICT4what?-Using the choice framework to operationalise the capability approach to development. J Int Dev 22(5):108–117
May J, Barrantes R (2015) Impact of research or research on impact: more than a matter of semantics and sequence. In: Chib A, May J, Barrantes R (eds) Impact of information society research in the global South. Springer, Heidelberg, pp 283–291. https://doi.org/10.1007/978-1-4419-7931-5
Misuraca G, Centeno C, Torrecillas C (2014) Measuring the impact of e-inclusion actors: impact assessment framework main report. European Commission, Luxembourg
Njenga J (2018) Digital literacy: the quest of an inclusive definition. Read Writ 9(1):1–7
Nulty DD (2008) The adequacy of response rates to online and paper surveys: what can be done? Assess Eval High Educ 33(3):301–314
Radovanović D, Holst C, Banerjee Belur S, Srivastava R, Vivien Houngbonon G, Le Quentrec E et al (2020) Digital literacy key performance indicators for sustainable development. Soc Incl 8(2):151–167. https://doi.org/10.17645/si.v8i2.2587
Research ICT Africa (2020) Digital Futures: South Africa's Readiness for the 4IR [Policy Paper]. Research ICT Africa. https://researchictafrica.net/publication/digital-futures-south-africas-readiness-for-the-4ir/
Scheerder A, Van Deursen A, Van Dijk J (2017) Determinants of Internet skills, uses and outcomes. A systematic review of the second-and third-level digital divide. Telematics Inform 34(8):1607–1624. https://doi.org/10.1016/j.tele.2017.07.007
Sen A (1999) Development as freedom, 1st edn. Oxford University Press, Oxford
StatsSA (2022) General household survey. Statistics South Africa, Pretoria
Uys C, Pather S (2016) Government public access centres (PACs): a beacon of hope for marginalised communities. J Community Inform 12(1):21–52
Uys C, Pather S (2020) A benefits framework for public access ICT4D programmes. Electron J Inform Syst Dev Ctries 86(2):1–18. https://doi.org/10.1002/isd2.12119
Van Audenhove L, Baelden D, Mariën I (2016) Quick-scan analysis of multiple case studies: SMIT policy method brief N°1. Studies on Media, Information and Telecommunication (SMIT), Brussels
Van Deursen AJAM, Van Dijk JAGM (2019) The first-level digital divide shifts from inequalities in physical access to inequalities in material access. New Media Soc 21(2):354–375. https://doi.org/10.1177/1461444818797082
WHO (2016) Monitoring and evaluating digital health interventions: a practical guide to conducting research and assessment. World Health Organisation, Geneva. Available at: https://apps.who.int/iris/handle/10665/252183
World Bank (2021) The World Bank in South Africa. Available at: https://www.worldbank.org/en/country/southafrica/overview#1

Chapter 13
Digital Health Literacy—A Prerequisite Competency for the Health Workforce to Improve Health Indicators in Times of COVID: A Case Study from Uttar Pradesh, India

Ritu Srivastava and Sushant Sonar

Introduction

The COVID-19 pandemic, its continuous waves, and the following lockdown have brought upon the entire world to a standstill. The pandemic raised some immediate questions on the capacity and access of the existing public healthcare system and the government relief measures. Furthermore, where the affluent sections of the society had access to smart devices and internet connection and were able to shift to online mode to access their basic needs and health services, whereas the other section of the society struggled to make their ends meet with no hope at sight.

The Uttar Pradesh state in India has had to address many challenges, unique to the complexities of the State giving a disproportionate proportion of global and India-specific burden of disease and deaths. The state developed, implemented, and drove state-wide adoption, of this comprehensive COVID-19 unified data platform, which has been able to bring together all (public and private) stakeholders engaged in the state's COVID-19 health response. The agility and layered architectural framework with which this integrated platform has been developed facilitates inter-linkage with modules developed for each key stakeholder, i.e., surveillance module for state/district surveillance teams and field tracking teams (via Case Tracking App), facility module for 1000+ facilities, and lab module for 350+ laboratories covering 98% COVID specific private and public facilities and labs in UP, to update

R. Srivastava (✉)
Senior Specialist, UPTSU, & Director, Jadeite Solutions, Delhi, India
e-mail: ritu.srivastava@ihat.in

S. Sonar
Senior Specialist – Data Visualisation, UPTSU – IHAT, Mumbai, India
e-mail: sushant.sonar@ihat.in

D. Radovanović (ed.), *Digital Literacy and Inclusion*,
https://doi.org/10.1007/978-3-031-30808-6_13

information, review case wise progress and take prompt action, as required (IHAT 2020).

Given the context of the current health systems and processes in public health infrastructure and human resources, digital health literacy played a crucial role in ensuring the preparedness of the healthcare system and mitigate its effects on individual and societal health. Digital health literacy entails knowledge, competence and skills of health workforce or an individual to attain, process, communicate and comprehend health information and services to promote and improve personal and community health services through effective health decisions (Broucke et al. 2020; Dunn 2019). Moreover, the adequacy of health literacy combined with digital literacy skills creating an enabling environment, implementing health policies, processes, systems, and health outcomes (Nguyen et al. 2020).

In recent years, digital health literacy has gained significant attention, owing to its association with social determinants of health. The WHO commission on Social Determinants of Health also recognized health literacy in determining health inequalities within low- and middle-income countries (CSDH 2008). The chapter brings the case study of UP's Integrated Unified Data Platform and the digital health literacy skills imparted to over health workforce and frontline workers. Using the digital health literacy instrument (DHLI), the study measures the improvement in digital health literacy skills and the competency of health work officials based on the contextual level of digital health literacy training imparted to them. The chapter aims to establish digital health literacy as a prerequisite requirement for health professionals and workforce for effective delivery of healthcare services specifically in public health sector. The chapter argues that digital health literacy skills are mandatory to ensure how health workforce and practitioners can integrate their knowledge and digital health literacy skills into optimal health behaviour.

The chapter is organised as follows: the literature review section investigates various digital health literacy measurements (including DHLI 1.0, DHLI 2.0, eHLF, etc) and establishes a linkage between digital health literacy and digital literacy. The methodology section defines different theories that have been applied for measuring digital literacy knowledge and proficiency skills which were further used to measure digital health literacy. The case study of GoUP on unified data platform for COVID-19 was presented followed by results and discussions. The findings section identifies that state and district health workforce who had digital literacy knowledge and proficiency skills and also received digital health literacy trainings was able to use mobile application effectively by entering the correct data as well as able to track the patients and analyse it for making informed decisions. The last section of this chapter provides a conclusion and recommendation that can be used for further research studies.

Literature Review

The previous section has established the relationship between digital literacy and health literacy to have competent digital health skills for any individual. This section covers digital literacy skills and competencies that are required for health officials and workforce to have for adopting digital health literacy skills through the literature review. This section covers different digital health literacy measurement instruments, including e-health literacy framework, DHLI 1.0, DHLI 2.0, and ehealth literacy scale (eHEALS).

Given issues of access to the internet and basic literacy, it is critical for health workforce and practitioners in public health settings to provide action-oriented health services or information that the recipient is able to benefit with the information or service. Digital literacy skills are critical for health workforce in the public health care with the emergence of patients involving digital healthcare services, for example, searching or looking health information online. Norman and Skinner in 2006 addressed the need of health workforce's digital competencies. Their 'Lily Model' for e-Health literacy defined digital health literacy as '*the ability to seek, find, understand and appraise health information from digital or electronic sources and apply the knowledge gained to addressing or solving a health problem*' (Norman and Skinner 2006). The Lily Model introduced eight-item eHealth Literacy Scale (eHEALS) instrument for measuring digital health literacy, comprising of three contextual literacies—(1) health literacy; (2) computer/digital literacy and science literacy and three analytical literacies—(1) traditional literacy, (2) information literacy, and (3) media literacy (Norman and Skinner 2006). Gilstad expanded the model by adding contextual, cultural, and social dimensions and defined it '*...ability to identify and define a health problem, to communicate, seek, understand, appraise and apply digital health information and welfare technologies in the cultural, social and situational frame and to use the knowledge critically in order to solve the health problem*' (Gilstad 2014).

ehealth literacy framework (eHLF) by a research group from Deakin University (Norgaard et al. 2015) consisted of seven dimensions derived from a structured mapping process involving healthcare workforce and patients. These seven dimensions are categorised as (i) ability to process information, (ii) engagement in own health, (iii) ability to actively engage with digital health services; (iv) feel safe and control; (v) motivated to engage with digital services; (vi) access to digital services that work, and (vii) digital services that suit individual needs (Norgaard et al. 2015) together with a multifaceted understanding of digital health literacy. Subsequently, the ehealth literacy questionnaire (eHLQ) has been developed based on the aforementioned seven dimensions.

Other theories by Koopman et al. (2014) in their PRE-HIT instrument included '*readiness for health information technology*' predicting the use of health information from patient's point of view. Different models—eHLF, eHLQ and eHLA have been developed in combination of health literacy and digital literacy with self-assessment elements for screening purposes in projects involving health solutions.

Digital Health Literacy Instrument (DHLI) was introduced in 2017 having a similar approach of combining skill and self-assessment (Van der Vaart 2017). DHLI consisted of 21 self-assessed items supplemented with seven performance tasks that focuses on handling digital information, primarily related to using health services through internet and ability to connect health professionals. DHLI included both Health 1.0 tools and Health 2.0 tools promoting the feasibility of assessment that is done with self-reportage of healthcare consumers' perceived skills. To measure the ability to use a broad spectrum of DHLI1.0 and DHLI2.0 skills, diverse range of digital literacy skills are important for retrieving health information alone for health workforce. DHLI 1.0 defines any health professional or practitioner who first need to have operational and navigational skills to use digital devices, including able to use touch screen, keyboard and search information on the internet. Second, they have the ability to evaluate skills to search, appraise and apply online information. Whereas to use Health2.0 applications, health workforce needs additional skills related to interactivity on the web. This includes adding self-generated content to the internet and be able to understand own and others' privacy and consent (Norman 2011; Van der Vaart 2017). The literature explains having necessary digital literacy and cooperating with digital services has become a critical skill for health professionals and public health workforce.

Relationship Between Digital Literacy and Digital Health Literacy Skills

The section also addresses the definitions of digital health literacy and preliminary digital literacy skills required to obtain the digital health skills in the current technological surroundings.

Globally, health literacy has been identified as a public health goal for facilitating healthcare services (Rootmanm 2003; Nutbeam 2000, JAMA 1999). Originally, health literacy has been conceived of as the '*...the degree to which individuals have the capacity to obtain, process and understand basic health information and services needed to make appropriate health decisions*' (Sentell et al. 2021). Scholars have argued the limitation on the conceptualisation of health literacy, maintaining that a focus should be put on the ability to factually contribute to promoting healthcare services risk prevention services in light of the healthcare system's demand and complexity (Pleasant et.al, 2016). Based on these theories, health literacy is a combination of two components— (1) individual health literacy, i.e. any individual ability to use health-related information to navigate the healthcare service system (Ancker et al. 2020) and (2) organisational health literacy, i.e. health officials and workforce capability to understand health information (Palumbo 2016).

Digital health literacy emerges as a new concept that is defined as '... *an extension of health literacy within the context of technology or electronic sources of information to understand and address any health problems as defined in* Fig. 13.1 (Zakar et al. 2021)'. In particular, digital health literacy entails the ability to access and use ICT and digital tools to co-design and/or co-deliver services intended for promoting health information, preventing risks, and contributing to collective health well-being (Azzopardi-Muscat and Sørensen 2019).

The digital literacy skills include ability to use and navigate the internet for their purpose and able to retrieve the information for analytical and making informed decisions (Radovanović et al. 2020). The intersection of digital literacy and health literacy has been described by multiple research factors that are related to the

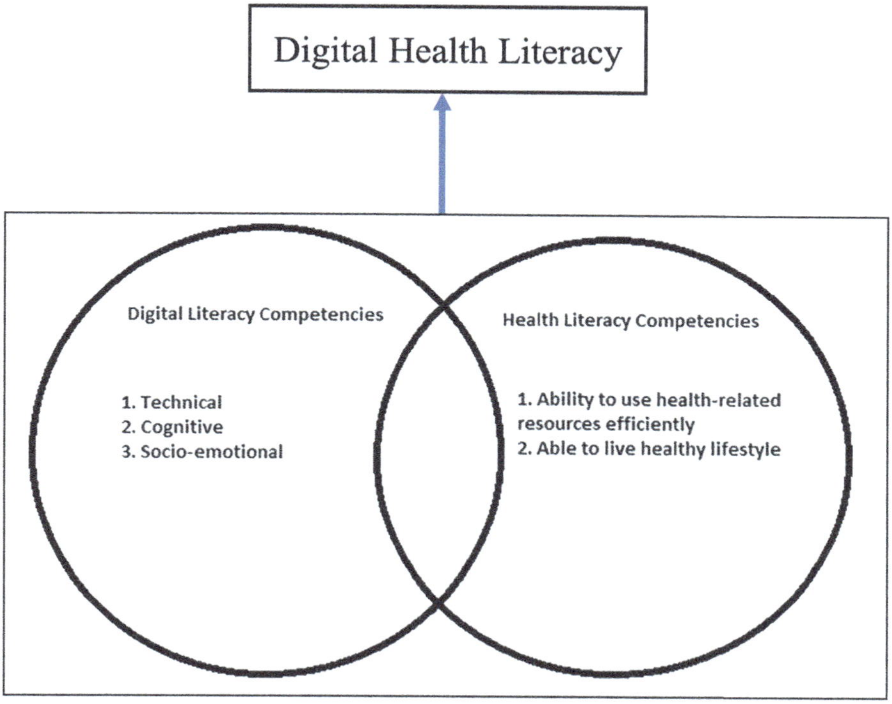

Fig. 13.1 Digital literacy vs. health literacy

acceptance of technological solutions for health management purposes. Most of these research factors have emphasized on the usefulness, ease of use, social factors, effectiveness, facilitating conditions, performing and effort expectancy of a given technology in an everyday context (Sohn and Kwon 2020).

Digital literacy in many theories has been defined as '*…the ability to understand and use information in multiple formats from a wide variety of sources when it presented via digitally*' (Eshet-alkalai 2004) (Gilster 1997). Gapski sub-divides digital literacy into two main strands (1) instrumental technological; and (2) normative media educational referring to usage/functionality of the technology (Gapski 2007). These strands are further described in three forms of competencies (a) interpreting messages, (b) selecting/choosing messages, and (c) articulating messages (Pietrass 2007).

In context of digital health technology usage, these models have not considered digital literacy as a tool to determine individual's intention towards the adoption of digital health technologies for health managing purposes. These competencies further measure functional, cognitive, and socio-emotional proficiencies of any individual. The cognitive dimension refers to the individual's ability to understand how to use and produce digital sources. The socio-emotional dimension refers to an individual's skills in using technology responsibly and has specific usefulness. Several case stories have identified if any individual has been exposed to technology and the use of digital health devices, digital learning or digital literacy happens automatically (Wang et al. 2013). All these aspects of digital literacy are fundamentally related to critical digital literacy or to the information that is produced or created by someone on the digital platform for a specific purpose and it is accessed or assessed by people.

Therefore, using any digital device effectively, users must or acquire the ability or 'inherent digital literacy' skills to use digital health technologies efficiently (Webber and Johnston 2017). This raises a recognition that digital literacy as an inherent skill to understand the impact of social and environmental conditions on the digital health literacy. In a practical sense, digital health literacy skilled health workforce is able to do the following:

1. Understand the potential of health services to support in doing their job better
2. Use various digital technological tools to
    a. Access and understand online health information in a range of multimedia formats
    b. Search, prioritise and bookmark relevant online resources
    c. Search for, learn, form and participate in online communities whether one to one or one to many or many to many
    d. Reproduce and share existing online health resources to build others understanding
3. Evaluate the appropriateness of what they find for their community

For each of these skills, individuals may have beginner, intermediate or advanced abilities.

Methodology

The study has applied cross-sectional method to measure digital health literacy of health workforce and practitioners. The literature has established the connection between digital literacy knowledge and skills and digital health literacy. In this paper, the digital health literacy training parameters are defined based on the key responsibilities given to state and district health officials and workforce (including frontline workers). The training on digital health literacy was imparted to the state health officials, workforce and frontline workers.

For this purpose, authors have used Aviram and Eshet-Alkalai (2006) theory for measuring digital literacy knowledge and for digital literacy proficiency skills and used Lynch and Swing (2006) theory for measuring the knowledge and proficiency skills digital literacy knowledge of stakeholders is defined on the level of their stakeholders' ability to use digital devices. Authors used observational method for measuring existing digital skills and proficiency levels of stakeholders. Observational method was applied to 50 stakeholders (trainees) representing state- and district-level health officials. Based on observational method, authors have categorized the digital literacy knowledge and proficiency level (Table 13.1) of trainees based on the knowledge of using digital devices and web and mobile applications.

Table 13.1 Categorisation of digital literacy knowledge and proficiency level

#	Stakeholders (trainees)	Responsibility	State/ district	Digital literacy knowledge level[a]	Digital skills – proficiency level[b]
1	State Surveillance Officer (SSO)	Monitoring of COVID platform. From admin perspective, s/ he can oversee from DSO to Death Committee users' work	State	DLK4	DLP5
2	Chief Medical Officer (CMO)	Monitoring & maintaining repository for surveillance cases, sampling, facilities transaction	District	DLK4	DLP5
3	District Surveillance Officer (DSO)	Monitoring repository for surveillance cases, sampling, facilities transaction	District	DLK4	DLP5
4	Epidemiologist	Monitoring & maintaining repository for surveillance cases, sampling, facilities transaction	District	DLK4	DLP5
5	Data Manager	Monitoring & maintaining repository for surveillance cases, sampling, facilities transaction	District	DLK3	DLP3
6	RRT (Rapid Response Team)	Follow-up of surveillance cases, contact tracing and home eligibility check for home isolation	District/ Block	DLK3	DLP3
7	Block Community Process Manager (BCPM)	Follow-up of surveillance cases, contact tracing and home eligibility check for home isolation	Block	DLK3	DLP3
8	Health Facilities (public & private) hospitals	Updating day-to-day transaction of positive cases	District/ Block	DLK2	DLP2
9	Lab admin and staff members	Uploading the result, maintaining CT value	District	DLK2	DLP2
10	Death committee users	Auditing of death [a]11 Medical college staff members are designated as death committees	State	DLK3	DLP3

[a]Digital literacy knowledge defined on the basis of literature review. Digital literacy knowledge is defined in Table 13.2

[b]Digital literacy skill proficiency is defined on the basis of different proficiency skills

Table 13.2 Literature review on digital literacy knowledge level

Digital literacy sub-discipline	Ability to do	Literature	DL knowledge level
Computer literacy	An understanding of how to use computers /smartphone and application software for practical purposes	Martin and Grudziecki (2006)	DKL1
Technology literacy	Computer skills and the ability to use computers and other technology to improve learning, productivity, and performance	U.S. Department of Education (1996)	
Information literacy	Finding and locating sources, analysing and synthesizing the material, evaluating the credibility of the source, using and citing ethically and legally, focusing topicsand formulating questions in an accurate, effective, and efficient manner	Eisenberg, Lowe, and Spitzer, in Meyer et al. (2008, p. 2)	DKL2
Media literacy	A series of communication competencies, including the ability to access, analyze, evaluate and communicate information in a variety of forms including print and non-print messages	Alliance for a Media Literate America (2010)	
Communication Literacy	Learners must be able to communicate effectively as individuals and work collaboratively in groups, using publishing technologies (word processor, database, spreadsheet, drawing tools...), the Internet, as well as other electronic and telecommunication tools	Winnepeg School Division (2010)	DKL3
Visual literacy	The ability to 'read,' interpret, and understand information presented in pictorial or graphic images; the ability to turn information of all types into pictures, graphics, or forms that help communicate the information;	Stokes (2002)	DKL4

To measure digital health literacy skills, authors have used abbreviated version of the DHLI 2.0 used by the global COVID-HL Consortium. The literature review section identifies that digital health literacy skills play a vital role in accessing and using digital health services to seek health-related information and make informed decisions (Xie et al., 2020). The literature also proves that if digital health literacy training is imparted effectively to health workforce and practitioners, the use of digital health platform is further improved, and collection of data is recorded accurately. In order to measure the effectiveness of the digital health literacy training, the paper attempts to evaluate the pre-and-post training knowledge and skill improvement of trainees.

Post training, the paper measures advancement in trainees' digital health literacy skills using DHLI 2.0 framework based on their ability to access the platform in different modules. To measure the effectiveness of the digital health literacy training, the number of COVID-positive and probable cases captured by different stakeholders/trainees (health workforce and practitioners) on the platform. The data on Unified COVID platform is recorded from May 2020 to September 2021 for three modules—Surveillance, Laboratory and Facility.

Limitations

As the Unified data COVID platform was developed and implemented in the span of two months and rolled out and implemented across the state in 2020 therefore, the paper did not conduct the baseline study to understand the existing basic digital literacy skills and digital health literacy of health workforce. Moreover, the training was imparted to over 200,000 health workforces, therefore mapping out the existing digital health literacy skills was not possible for this paper. Therefore, for the benefit of this study authors focused on end-line usage of the application (mobile and web) and applied observation method to selected trainees for measuring the effectiveness of the training.

COVID-19 and Digital Health Services in Uttar Pradesh

The COVID-19 continues to impact lives across the world, hampering health equity and creating socio-economic growth of countries. The healthcare facilities were overwhelmed and finding it difficult to manage the sudden demand in hospital beds, medicines and supplies.

This pandemic has exposed the vulnerabilities of our healthcare system, but at the same time, it also led to unprecedented growth towards adopting digital technologies in healthcare systems and management for delivering efficient healthcare services despite maintaining social distancing. Preparing for this crisis, the Government of India made significant advancements in the way digital healthcare systems and innovative digital solutions adopted and developed for effective healthcare service delivery. In response to COVID-19 outbreak, along with the state governments and private stakeholders, the Indian government took immediate and necessary actions to tackle the pandemic by setting up dedicated COVID-19 hospitals, isolation centres and tech-enabled mapping of resources. Within a span of few months, the Government of India initiated a platform at scale and end-to-end workflow-based solutions.

While health being a state subject, many states in India initiated development and deployment of new digital tools and digital services as per their citizen needs. In an effort to deal with the COVID-19 pandemic holistically, Uttar Pradesh (UP) was one of the first states to develop mobile applications and digital information systems, moving towards developing an end-to-end integrated surveillance platform to monitor and track COVID-19 patients.

Over 235 million people live in Uttar Pradesh living in 107688 villages distributed in 75 districts. In terms of the public health infrastructure, there are 25812 public health facilities comprising 141 district hospitals (DH), 278 special hospitals (SH), 943 community health centres (CHCs), 3602 public health centres (PHCs) and 20848 sub-centres, according to UP Ke Swasthya Kendra portal.

The COVID-19 pandemic has disproportionately affected individuals, putting them at the risk of increased morbidity and mortality, underscoring the urgent need to provide basic healthcare services and maintain global good health and well-being for achieving SDG goals (The Global Goals). Moreover, 5 million migrant workers which were making back to their home in May 2020 (The Hindu 2020), it became crucial for the UP government to take necessary measurements not only to ensure that migrant workers reach home safely but also to provide relevant information and health services to people living in rural areas.

Keeping the size and scale of the state in mind, UP Government developed a COVID-19 Unified Data Platform as a single source of information and end-to-end case management platform across the continuum of care for COVID-19. The platform was kicked off on March 2020 and was stabilized by May 2020 and has been leveraged across both the waves for effective management of the pandemic.

A Unified Data Platform for COVID-19 – A Case Study of Uttar Pradesh

This section covers the case study of Unified Data Platform development, implementation and statewide adoption through various capacity building and digital health training on the platform to its 200,000 state and district health officials spread across 75 districts and over 59000 village councils of the state. These health officials included facility staff members, laboratories and district-level health nodal officials, surveillance team and helpline coordinators.

The Unified COVID-19 platform was designed to facilitate end-to-end case management of patients and health workers digitally. The mobile-and web-based data platform gives real-time view of the status of the cases across local districts by collecting and aggregating data from its tracking and contact modules (Fig. 13.2).

The surveillance of COVID-19 patients was one of the prime aspect of this Unified platform. It included the steps from case registration, case assigning, tracking application, assigning the test type and lab, lab module, allocation of facility or home isolation in case of positive case, case update at the facility level, referral to other facility, closure of the case to handling of death. It included the following human resources for capturing the data systematically:

1. Case registration by SSO, DSO, lab or field team
2. Case is assigned for in-person verification by the DSO to the tracking team
3. Reviewing and verifying the case details by tracking team through the case tracking application

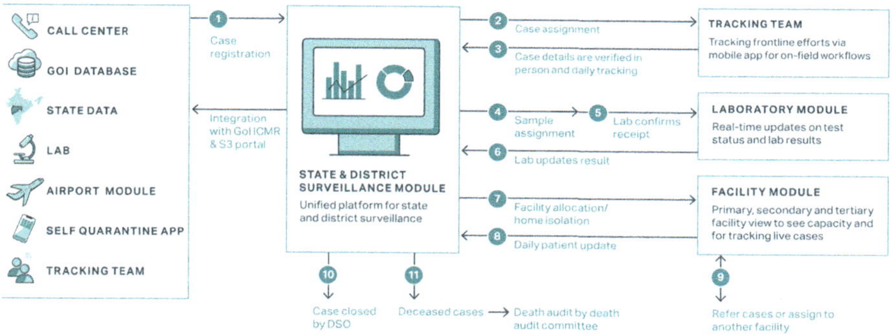

Fig. 13.2 An integrated work-flow based system for COVID cases management (Source: Redesigned from IHAT, the Uttar Pradesh COVID-19 Unified Data Platform)

4. DSO assigns the test type and lab
5. Lab confirms sample delivery and receipt
6. Lab confirms and updates results
7. In case of positive case identified, the DSO allocates a facility or home isolation
8. Facility user updates case details daily
9. Referral to other facility
10. Case closure by DSO
11. Handling of death cases

Based on the work-flow based system and following brick-mortar model, the platform was designed to assist the following stakeholders. The digital health literacy training was structured around key features of Unified Data Platform for COVID-19 modules (Table 13.3).

The platform architecture was designed to enable clear, immediate and easy coordination and referral linkages among state and district surveillance teams, field-tracking teams, laboratories and state's 235 million inhabitants.

Capacity Building and Training Method

The Unified COVID-19 data platform (mobile- and web-based) was ready to be rolled out in mid-March 2020, therefore, one of the tasks is to roll out state-wide training and capacity-building sessions for health workforce (including doctors, laboratory staff members, facility staff members) and frontline workers (including ANMs and ASHAs). The Directorate of Medical and Health Services (DGMH), GoUP adopted a structured approach for the training of health officials) and front-line workers.

The digital health literacy training on COVID cases management was given at two levels—(1) facility level and (2) community-level trainings. Facility-level trainings were given to 714 medical teams, 12051 doctors, 12983 staff nurses and 43,140

Table 13.3 Key stakeholders and key features

Key stakeholders	Key features
Policy and decision-makers	1. Single source for all data and analytics across stakeholder groups 2. Integrated dashboard for swift decision making
State/district surveillance teams	1. Singular point of case registration with end-to-end case management and post discharge case follow up 2. Integration of multiple data sources (lab, facility, field, citizen, call centre)
Lab team	1. System team generated real-time receipt status updates for samples and consignments 2. Ability to upload the results at the lab – updated real time at district/facility/field team level
Facility team	1. Medical and epidemiology record maintained against the unique case ID across the case life cycle 2. Seamless inter-facility referral
Field team	1. Integrated with real time updates to and from state/district and lab team 2. In-person and verification and daily follow up of each registered case 3. Contact tracing on the field 4. Collaboration with all field teams, facility, lab and state/district team
State residents	1. Direct beneficiary engagement through advisory messages for registered users 2. Geo-fencing and movement alerts 3. Self-assessment, tracking and helpline support 4. Self-registration of passengers traveling to UP via air 5. Easy access to COVID lab test results via an online single platform for the public across facility and test types 6. Access and search COVID test and collection centre details

Source: Restructured based on the work-flow designed for the Unified data COVID application

paramedical staff whereas community-level training was imparted to 589 district officials (including CMO, DSO), 3138 block officials (including BCPM) and further to health frontline workers (IHAT, Unified Data Platform). The below Fig. 13.3a gives the overview of training method and the use of different technologies for conducting trainings.

DGMH, GoUP conducted digital health literacy training(s) on using Unified COVID-19 Data platform. On the basis of platform workflow, the state and district health workforce were given different responsibilities for the management of COVID-19 cases and probable cases. The digital health literacy training was structured around different modules and the key responsibilities for managing the platform given to trainees. These trainings were conducted through various online channels such as Zoom in small groups or the advent of remote training via different mediums as referred in Fig. 13.3b.

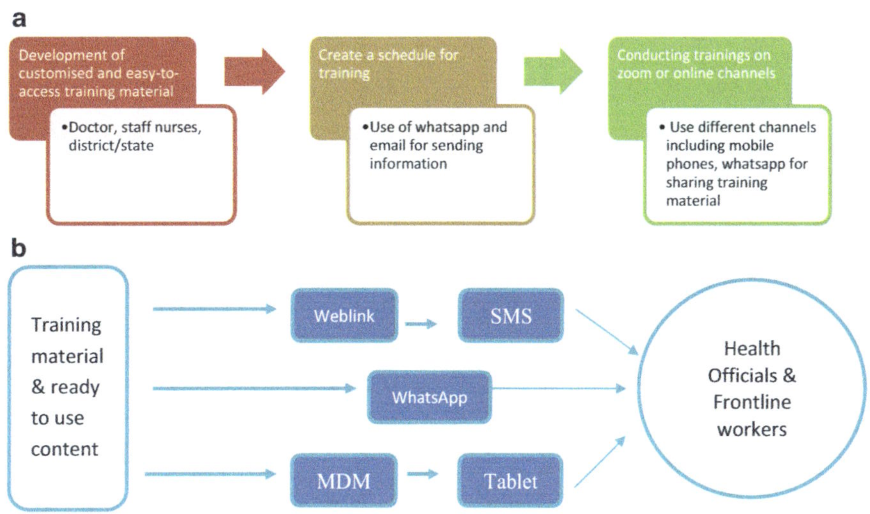

Fig. 13.3 (**a**) Training process flow. (**b**) Use of technology for training

Results and Discussion

This section of the paper analyses digital health literacy training based on different modules of the platform aligning with DHL2.0 skills and also measures the improvement in their digital health skills if trainees (health workforce and practitioners) have already digital literacy knowledge and proficiency skills. The effectiveness of the digital health literacy training is measured based on the number of recorded COVID-19 positive and probable cases captured by different stakeholders/trainees (health workforce and practitioners) on the digital mobile-and-web platform.

The training on digital health literacy was imparted in the phase-1 of COVID-19 pandemic. Health workforce and practitioners working at the district and below levels, have basic literacy knowledge, including ability to use digital devices, ability to navigate and search information. In addition to leveraging Zoom as a platform to train the state, district, RRT, facility and lab teams on the end-to-end surveillance module short, easy-to-view videos were developed and disseminated for each module.

At the state level, SSO, CMO and DSO at district level have received training on all four modules for monitoring and surveillance of the platform (Table 13.4). DSO acts as a nodal officer at district to monitor and analyse surveillance cases (COVID positive and probable cases). Whereas below district level, facility staff members of public health centre received training on facility module. Frontline workers (ANMs) were proactively informed and updated about new content material available on the website via SMS reminders and updates. This helped officials and health providers across the state to become more confident to use the platform.

The surveillance module is larger than other modules, including facility, laboratory and other modules as per the architecture of the platform. The training on digital health literacy was imparted in the phase 1 of COVID-19 pandemic and support to record the data continued. Health workforce and practitioners working at district and below levels have basic literacy knowledge, including ability to use digital devices, ability to navigate and search information.

At the district level, DSOs received training on all three modules and have the capability to monitor and analyse surveillance cases. Whereas below district level health workforce, facility staff members of public health centre received training on facility module.

The data in Table 13.5 shows that during the peak time of COVID-19 phase 1 (2020) and phase 2 (2021) health workforce and practitioners are able to record the data. District- and block-level health workforce, including labs, public health centres located in rural areas are able to enter data of cases on the platform. It reflects that district- and block-level health workforce and officials who have received digital health literacy training and continuous support through other digital channels, including helpdesk are able to enter the record of cases on the ground.

On a macro level, decision-makers often use specific, quantifiable markets to guide the commissioning of new digital health services and strategies. Collection of

Table 13.4 Categorisation of trainees on digital health literacy modules

State and district level	Digital health literacy training on modules of COVID platform				Improvement in digital health	Digital health literacy Scale[a]
Stakeholders	Surveillance module	Lab module	Facilities module	Death module		
State Surveillance Officer (SSO)	√	√	√	√	1. Ability to monitor the cases effectively 2. Able to analyse the work of DSO to Death Committee work 3. Able to analyse positive and probable cases and protect patient's information	DHL6
Chief Medical Officer (CMO)	√	√	√	√	1. Improvement in maintaining COVID-19 repository cases for surveillance purpose 2. Able to visualise and analyse the number of sampling required for efficient monitoring 3. Able to visualise and analyse number of additional facilities required for monitoring of COVID-19 cases	DHL6
District Surveillance Officer (DSO)	√	√	√	√	1. Knowledge of monitor and analysis surveillance cases 2. Able to transfer cases to referral facility	DHLI 5

(continued)

Table 13.4 (continued)

State and district level	Digital health literacy training on modules of COVID platform				Improvement in digital health	Digital health literacy Scale[a]
Stakeholders	Surveillance module	Lab module	Facilities module	Death module		
Epidemiologist	√	×	×	×	1. Improvement in maintaining COVID-19 repository cases for surveillance purpose 2. Able to visualise and analyse the number of sampling required for efficient monitoring 3. Able to visualise and analyse number of additional facilities required for monitoring of COVID-19 cases	DHLI 5
Data Manager	√	×	×	×	1. Knowledge of COVID portal 2. Knowledge to retrieve information from COVID portal 3. Team alignment w.r.t COVID portal 4. Able to identify case and transfer the case to specific facility as per case requirement 5. Ability to evaluate cases 6. Able to create analytical reports 7. Able to communicate with state team for the addition of facility and inactive of private facility	DHLI 4

(continued)

Table 13.4 (continued)

State and district level	Digital health literacy training on modules of COVID platform				Improvement in digital health	Digital health literacy Scale[a]
Stakeholders	Surveillance module	Lab module	Facilities module	Death module		
RRT (Rapid Response Team) & BCPM (Block Community Process Manager)	√	×	×	×	1. Able to handle assigned cases 2. Able to identify cases w.r.t home-isolation or required facility 3. Communicate to DSO for providing facility	DHLI 4
Health Facilities (public & private) hospitals	×	×	√	√	1. Able to enter data of positive cases on the portal 2. Able to monitor day-to-day transaction of cases	DHLI 3
Lab admin and staff members	×	√	×	×	1. Able to enter the lab result on digital device 2. Able to modify and or update the case result	DHLI 3
Death committee users	×	×	×	√	1. Able to upload data on the portal regarding case result 2. Able to identify death of patients due to COVID	DHLI 3

[a]Digital health literacy dimensions: The table is reproduction of the digital health literacy scale and its measurement in consultation with stakeholder's knowledge and skill improvement post-training. The definition of DHLI on different dimensions of using digital skills for their work purpose

accurate record helped decision-makers to make decisions on imposing lockdown, necessary action for vaccination drive and development of new facilities.

Therefore, if digital literacy skills and proficiency become an integral part of digital health literacy for all health professionals, including health front-line workers it will be easy for the public health sector to deal with such pandemic. It also emphasised that there is a need to undertake specialised leadership training for programme roles similar to the SSO and DSO in order to equip the Government officials, to lead any programme intervention end-to-end at the state/district level.

Table 13.5 No of cases recorded on Integrated Unified Data Platform on COVID

	Surveillance module	Lab module	Facilities module
Month-year	Total surveillance cases	Total sample tested	Total number of cases registered in facility
May-20	340,724	242,127	7981
Jun-20	523,867	543,781	19,293
Jul-20	1,624,329	1,947,410	55,061
Aug-20	2,926,060	3,645,869	62,901
Sep-20	3,693,786	4,755,498	59,175
Oct-20	3,646,693	4,912,499	29,827
Nov-20	3,312,061	4,661,428	23,247
Dec-20	3,285,872	4,672,275	13,170
Jan-21	2,744,440	3,922,135	4646
Feb-21	2,430,247	3,477,913	1332
Mar-21	2,618,294	3,688,435	4839
Apr-21	5,077,220	6,872,381	68,288
May-21	6,292,801	8,497,322	41,093
Jun-21	5,631,871	8,229,408	2322
Jul-21	5,200,770	7,677,149	503
Aug-21	4,596,422	6,762,318	241
Sep-21	4,197,227	6,217,581	132

Data Source: *UPCovidtracks.in*

Conclusion

Digitally enabled competent health workforce improve in maintaining health records and strengthen health systems and able to meet the challenges of responding to the changing health needs of the public. Digital health literacy (DHL) is crucial for health workforce as it enabled them not only to ensure patient safety but to monitor and track them and enable them to refer nearby lab and healthcare facilities.

Moreover, health workforce can be relieved from time-consuming routine tasks and interact better with patients (OECD, 2019). It has been emphasised that digital transformation in health sector is much more than going paperless but it is linking the existing database or digitalising existing tasks. Digital transformation means that the various digital technologies need to be leveraged to design appropriate, effective and efficient models of care and delivering health services. Hence, digitally skilled health practitioners are able to adapt these models effectively and use data to improve health service delivery—through, for example designing structured and tailored or better-coordinated health services—remain rare.

The digital health literacy training(s) should focus on different skills and competencies. These training(s) should have a tailored approach required for particular healthcare worker group, role, level of seniority and different geographical (including urban, semi-urban and rural) setting. Moreover, these digital health literacy framework and training(s) should be regularly updated with novel digital health technologies, to be applicable to low-and middle-income countries. Emerging economies like India need to have a dedicated digital health team especially focusing on rural population.

The case study has showcased that if digital health literacy training is imparted structurally, it helps government officials to target health workforce with specific responsibilities to monitor and track emergency health crisis. The prevalent competency domains identified represent essential inter-professional skills to be incorporated into healthcare workers' training.

Acknowledgement Authors acknowledge our colleague, Specialist, Mr. Vishan Sharma, who helped us with collection of data and given its insights for the case study.

References

Ad Hoc Committee on Health Literacy for the Council on Scientific Affairs, American Medical Association (1999) Health literacy: report of the Council on Scientific Affairs. JAMA 281(6):552–557. [Medline: 99144710]. https://doi.org/10.1001/jama.281.6.545

Ancker J, Grossman L, Benda N (2020) Health literacy 2030: is it time to redefine the term? J Gen Intern Med 35(1):2427–2430

Aviram A, Eshet-Alkalai Y (2006) Towards a theory of digital literacy: three scenarios for the next steps. Eur J Open Dist E-Learn 9

Azzopardi-Muscat N, Sørensen K (2019) Towards an equitable digital public health era: promoting equity through a health literacy perspective. Eur J Public Health 29(3):13–17

Broucke SVD, Levin-Zamir D, Schaeffer D, Pettersen K, Guttersrud Ø, Finbråten H, de Arriaga MT, Vrdelja M, Link T, Pelikan J (2020) Digital health literacy in general populations—an international comparison. Eur J Public Health 30:5

Census of India, 2011–2021; https://www.census2011.co.in/census/state/uttar+pradesh.html

Commission on Social Determinants of Health (CSDH) (2008) Closing the gap in a generation: health equity through action on the social determinants of health. Final Report of the Commission on SDH; World Health Organization, Geneva

Dunn P, Hazzard E (2019) Technology approaches to digital health literacy. Int J Cardiol 293:294–296

Eshet-alkalai Y (2004) Digital literacy: a conceptual framework for survival skills in the digital era. Pool

Gapski, H. (2007). Some reflections on digital literacy. Proceedings of the 3rd International Workshop on Digital Literacy (pp. 49-55). Crete CEUR-WS.org. http://ceur-ws.org/Vol-310/paper05.pdf

Gilstad H (2014) Toward a comprehensive model of eHealth literacy. In: Proceedings of the 2nd European Workshop on Practical Aspects of Health Information. Presented at: 2nd European Workshop on Practical Aspects of Health Information; 2014 May 19, Trondheim, pp 19–20. http://ceur-ws.org/Vol-1251/paper7.pdf

Gilster P, Glister P (1997) Digital literacy piercing purple want more papers like this?

IHAT (2020) The Uttar Pradesh COVID unified data platform, https://www.ihat.in/resources/the-uttar-pradesh-covid-19-unified-data-platform/

Koopman RJ, Petroski GF, Canfield SM, Stuppy JA, Mehr DR (2014) Development of the PRE-HIT instrument: Patient readiness to engage in health information technology. BMC Fam Pract 15(1):18

Lynch D, Swing S P (2006) ACGME outcome project from ACGME – the outcome project., http://www.acgme.org/outcome/assess/keyConsider.asp

Nguyen HC, Nguyen MH, Do BN, Tran CQ, Nguyen TT, Pham KM, Pham LV, Tran KV, Duong TT, Tran TV et al (2020) People with suspected COVID-19 symptoms were more likely depressed and had lower health-related quality of life: The potential benefit of health literacy. J Clin Med 9:965

Norgaard O, Furstrand D, Klokker L, Karnoe A, Batterham R, Kayser L et al (2015) The e-health literacy framework: a conceptual framework for characterizing e-health users and their interaction with e-health systems. Knowl Manag E-Learn 7(4):522–540

Norman C (2011) eHealth literacy 2.0: problems and opportunities with an evolving concept. J Med Internet Res 13(4):e125. https://doi.org/10.2196/jmir.2035. [Medline: 22193243]

Norman CD, Skinner HA (2006) eHealth literacy: essential skills for consumer health in a networked world. J Med Internet Res 8(2):e9. https://doi.org/10.2196/jmir.8.2.e9. [Medline: 16867972]

Nutbeam D (2000) Health literacy as a public health goal: a challenge for contemporary health education and communication strategies into the 21st century. Health Promot Int 15(3):259–267. https://doi.org/10.1093/heapro/15.3.259

Palumbo R (2016) Designing health-literate health care organization: a literature review. Health Serv Manage Res 29(3):79–87

Pietrass M (2007) Digital literacy research from an international and comparative point of view. Res Comp Int Educ 2(1):1–12. https://doi.org/10.2304/rcie.2007.2.1.1

Pleasant A, Rudd R, O'Leary CP-O, Allen MA-L, Myers L, Parson K, Rosen S (2016) Considerations for a new definition of health literacy. National Academy of Medicine, Washington, DC

Radovanović et al (2020) Digital literacy key performance indicators for sustainable development. Social Inclusion 8(2):151–167. https://doi.org/10.17645/si.v8i2.2587

Rootman I (2003) Literacy and health in Canada: is it really a problem? Can J Public Health 94(6):405–412. [Medline:23061744]

Sentell T, Foss-Durant A, Patil U, Taira D, Paasche-Orlow MK, Trinacty C (2021) Organizational health literacy: opportunities for patient-centered care in the wake of COVID19. Qual Manag Health Care 30(1):49–60

Sohn K, Kwon O (2020) 'Technology acceptance theories and factors influencing artificial Intelligence-based intelligent products', Telemat. Informatics. https://doi.org/10.1016/j.tele.2019.101324

The Hindu (2020) Coronavirus | Uttar Pradesh to bring back migrant workers in phases, https://www.thehindu.com/news/national/other-states/uttar-pradesh-to-bring-back-migrant-workers-in-phases/article31423150.ece. Accessed on 17th Feb 2022

Van der Vaart R., Drossaert C. Development of the digital health literacy instrument: measuring a broad spectrum of Health 1.0 and Health 2.0 skills. J Med Internet Res 2017;19:e27. https://doi.org/10.2196/jmir.6709 https://www.ncbi.nlm.nih.gov/pmc/articles/PMC5358017/

Wang Q, Myers MD, Sundaram D (2013) Digital natives and digital immigrants: towards a model of digital fluency. Bus Inf Syst Eng 5(6):409–419. https://doi.org/10.1007/s12599-013-0296-y

Webber S, Johnston B (2017) Information literacy: Conceptions, context and the formation of a discipline. J Inf Lit 11(1):156–183. https://doi.org/10.11645/11.1.2205

Xie X, Xue Q, Zhou Y, et al. (2020) Mental Health Status Among Children in Home Confinement During the Coronavirus Disease 2019 Outbreak in Hubei Province, China. JAMA Pediatr 174(9):898–900. https://doi.org/10.1001/jamapediatrics.2020.1619

Zakar R, Iqbal S, Zakar M, Fischer F (2021) COVID-19 and health information seeking behavior: digital health literacy survey amongst university students in Pakistan. Int J Environ Public Health 18(8):4009–4029

Correction to: How Southeast Asia Can Better Arrange and Deliver Internet Policies So as to Defy the Digital Divide

Jason Hung

Correction to:
Chapter 5 in: Radovanović (Ed), *Digital Literacy and Inclusion*, https://doi.org/10.1007/978-3-031-30808-6_5

The book was mistakenly published with an incorrect affiliation for a Chapter 5 author. The original version has been revised.

Dr. Jason Hung's affiliation should be updated from:

"Department of Sociology, The University of Cambridge, Cambridge, MA, USA"

to:

"Department of Sociology, The University of Cambridge, Cambridge, England, UK."

The updated version of this chapter can be found at https://doi.org/10.1007/978-3-031-30808-6_5

D. Radovanović (ed.), *Digital Literacy and Inclusion*, https://doi.org/10.1007/978-3-031-30808-6_14

Afterword

Josef Noll

In conclusion, the book presents various aspects of digital inclusion, literacy, and transformation. It points out the need for "digital and mobile first" when developing content to reach every human. The challenge addressed is how to empower the actors in the ecosystem, for example, communities, schools, and governments, to gain the economic freedom to prioritize digital inclusion and digital literacy.

Indeed, the book is an eminent advocate for the needs and asks for a call to action for each of us. As an individual, how can you contribute to empowering your neighborhood, those who are left aside or those living in developing economies to get connected and digitally empowered? As a community leader, how can you contribute to the free access to information on the internet in selected places of your community, such as community learning and living labs, schools, libraries, primary health centers, and governmental buildings? As an industry, how can you contribute to programs connecting schools and communities? As a regional and national government, how can you contribute to digital connectivity and empowerment and support grassroots initiatives addressing digital literacy and inclusion?

For example, as promoted by the international GIGA project from UNICEF and ITU, school connectivity is an excellent entry point to see what is happening in every country. Similarly, grassroots projects like AHERI in Kenya, African Child Projects in Tanzania, the Zimbabwe Community Network, the Norwegian Basic Internet Foundation project in Africa, and many more are excellent entry points to join for a connected and empowered future.

J. Noll
University of Oslo, Oslo, Norway

D. Radovanović (ed.), *Digital Literacy and Inclusion*,
https://doi.org/10.1007/978-3-031-30808-6

Index

D. Radovanović (ed.), *Digital Literacy and Inclusion*,
https://doi.org/10.1007/978-3-031-30808-6

The manufacturer's authorised representative in the EU is Springer Nature Customer Service Centre GmbH, Europaplatz 3, 69115 Heidelberg, Germany. If you have any concerns regarding our products, please contact ProductSafety@springernature.com

Printed and bound by CPI Group (UK) Ltd, Croydon, CR0 4YY
28/09/2026
02224075-0006